# Multiphysics Modelling

# Multiphysics Modelling

## Materials, Components, and Systems


Murat Peksen

Multiphysics Energy Solutions (MES), Jülich, Germany

Academic Press is an imprint of Elsevier
125 London Wall, London EC2Y 5AS, United Kingdom
525 B Street, Suite 1650, San Diego, CA 92101, United States
50 Hampshire Street, 5th Floor, Cambridge, MA 02139, United States
The Boulevard, Langford Lane, Kidlington, Oxford OX5 1GB, United Kingdom



**British Library Cataloguing-in-Publication Data**
A catalogue record for this book is available from the British Library

**Library of Congress Cataloging-in-Publication Data**
A catalog record for this book is available from the Library of Congress

ISBN: 978-0-12-811824-5

For Information on all Academic Press publications
visit our website at https://www.elsevier.com/books-and-journals

*Publisher:* Matthew Deans
*Acquisition Editor:* Brian Guerin
*Editorial Project Manager:* Mariana L. Kuhl
*Production Project Manager:* R. Vijay Bharath
*Cover Designer:* Greg Harris

Typeset by MPS Limited, Chennai, India

# Dedication

Dedicated to my adored family and my precious students.

# The Very First Page

Es ist nicht das Wissen, sondern das Lernen, nicht das Besitzen, sondern das Erwerben, nicht das Dasein, sondern das Hinkommen, was den größten Genuß gewährt. Wenn ich eine Sache ganz ins Klare gebracht und erschöpft habe, so wende ich mich davon weg, um wieder ins Dunkle zu gehen, so sonderbar ist der nimmersatte Mensch, hat er ein Gebäude vollendet, so ist es nicht, um ruhig darin zu wohnen, sondern um ein anderes anzufangen.

***[It is not knowledge, but the act of learning, not possession but the act of getting there, which grants the greatest enjoyment. When I have clarified and exhausted a subject, then I turn away from it, in order to go into darkness again. The never-satisfied human is so odd; if he has completed a building, then it is not for living in it peacefully, but in order to begin another one.]***

***Johann Carl Friedrich Gauß (1777–1855)***
***in his letter to Farkas Wolfgang Bolyai (2 Sep. 1808)***

# Contents

# List of Figures

# List of Tables

# About the Author


## Murat Peksen

*Managing Director, Multiphysics Energy Solutions (MES), Germany*

Murat Peksen received his BSc degree from the Mechanical Engineering Faculty of the Istanbul Technical University (ITU), Turkey. This was followed by a MSc degree in mechanical engineering from RWTH Aachen University, Germany. He was a visiting scholar in the United States at the University of Massachusetts, and was conferred with a PhD in computational fluid dynamics from the Mechanical Engineering Department of Loughborough University, United Kingdom.

Dr Peksen has been engaged in multiphysics modelling research and education for more than 16 years. As an adjunct professor at Aachen University of Applied Sciences, Germany and a visiting professor at the University of Science and Technology of China, China, he has taught 'Multiphysics modelling of functional materials and components' – the methodology of modelling coupled physical interactions, simulating and analysing materials and components, for many years. He is responsible for the Modelling Master Class at the Joint European Summer School on Fuel Cells, Electrolysers, and Battery Technologies (JESS).

As a scientific head of the modelling and simulation activities, Dr Peksen served about 8 years for the Forschungszentrum Jülich GmbH (FZJ), Germany. He worked as a project manager in alternative power trains at the FEV Europe GmbH, Germany. He is also the founder and managing director of the 3D simulation assisted R&D and consultancy company Multiphysics Energy Solutions (MES), Germany.

Dr Peksen is author of numerous scientific publications and has been awarded from various research institutions worldwide.

# Preface

The use of computational continuum mechanics to predict simultaneously occurring physics has risen dramatically within the past years. In the early 1980s, this kind of solutions of complex phenomena by means of numerical analyses was mostly the field for academic people. Today the progress on the development of hardware and software in computer sciences together with the sophisticated numerical analysis facilities enable the use of various commercial codes by the scientific and industrial communities.

Especially, the capability to modify and implement user-specific demands makes them powerful and highly desired. However, their operation and handling several issues needs expertise in many engineering fields, as well as software and hardware knowledge. Moreover, to predict and evaluate the results of complex solutions that encompass multiple physical phenomena and interactions requires high level of understanding, interpretation skills of the user.

Within the last decade, I have held worldwide numerous lectures and seminars about multiphysics modelling and simulation for students, researchers, scientists and industrial personnel. During that time, I have recognised that there is still an insisting desire to ignore naturally coupled processes and the loss of invaluable information about the interaction among the elements building up a working system. However, all problems arising from natural phenomena are of multiphysics nature. Therefore all research is interrelated and it is not conceivable anymore to neglect this fundamental issue within the scientific and industrial communities, in order to capture the complexities of processes.

It is the purpose of this book to fill a gap in the available literature and provide the reader with a theoretical and practical knowledge of multiphysics modelling. Thereby, the skills required to analyse interacting engineering problems through numerical simulation are elucidated. The book comprises the methodology of modelling coupled physical interactions, simulating and analysing materials, components and systems. It is intended to be designed for graduate students, practicing engineers, scientist and industrial communities as an introductory and self-contained level, in order to shape the ability to see beyond boundaries – between technical disciplines.

The presented material is derived from interdisciplinary fields comprising computer science, numerical methods and their applications in energy, mechanical and many engineering fields based on my experience, gathered over the years.

As most of the contents are developed from my course notes and research, the book will also be valuable as a learning and teaching resource to support multiphysics modelling courses at undergraduate and graduate level. With its academic literature it will be a link between the academic world and professional engineers dedicated to R&D activities. My experience shows that a step by step developing approach has been a good methodology to successfully understand and solve complex multiphysics problems. Thus, I exerted myself to hold on this approach in this book, as well. Accordingly, the reader should first be familiar with important standalone physics, prior switching to the coupled behaviour of them. The topics will be enriched with practical examples throughout this book.

Working with students and scientists across the world has taught me much about educating the notion of multiphysics. I have tried in earnest to draw from this experience for the benefit of this book. Obviously, I am devoted to lifelong learning; therefore I welcome any suggestions and comments from you.

**Murat Peksen**

Aachen, December 2017

# Acknowledgements

*It has been well said by Shakespeare that 'Pupil Thy Work is Incomplete, Till Thee Thank the Lord and the Masters' and I sincerely believe this.*

In the name of Allāh, the Most Gracious, the Most Merciful;

I am deeply indebted to The Central Library of Forschungszentrum Jülich, Germany, for enabling me as an alumni to access their unique services in the fields of documentation, publication and information management. It goes without saying that this book would not have been possible without their support throughout the period of this work.

It has been a privilege to work closely with the team of Elsevier S&T Books to successfully complete this endeavour. My special thanks goes to Mr Brian Guerin and Mrs Mariana Kühl Leme for the whole editorial process.

I am profusely grateful to my worldwide scientific network for their in-depth discussions and fruitful opinions, providing a fascinating foundation for this book. I would like to express my heartfelt gratitude to Dr -Ing. Ali Al-Masri, for having always provided invaluable insights throughout the preparation of the work. My forever gratefulness to Prof. Dr Cevza Candan for her consistent encouragement and motivation during my whole academic career.

Finally, I wish to express my sincere appreciation to my mother, father, sisters and my beloved Martina for their unlimited support, love and sacrifice for making it possible to pursue the completion of this work. They always encouraged me, never trying to limit my aspirations that gave me the ability to see past boundaries – between countries, cultures or academic multidiscipline – has helped shape the foundation of this book.

Chapter 1

# Introduction to Multiphysics Modelling

## Chapter Outline



MULTIPHYSICS is the science, studying multiple interacting phenomena. With its complexity and covering a wide range of physical fields, multiphysics modelling is the royal discipline of numerical modelling to solve engineering problems. Numerical methods are the essential tool for predicting and simulating the multiphysical behaviour of interacting complex engineering systems. Therefore, multiphysics modelling has witnessed many significant developments in the last century, due to the progress in numerical methods.



Multiphysics Modelling. DOI: https://doi.org/10.1016/B978-0-12-811824-5.00001-8

The historical development of numerical methods show that they are attributed to astronomers and applied mathematicians who were concerned to transform their physical problem into a mathematical description, which they called *modelling*. This is the reason why most of the solvers of numerical codes are endorsed to the famous astronomers like Newton, Gauss or Euler etc. The fundamentals of numerical analyses i.e. algorithms-algebra, however, originate even to earlier times as it was for the first time used and named after another great scientific mind Abu Ja'far Muhammad ibn Musa al-Khwarizmi [1] who lived in the 8th Century during the Islamic empire in Todays Khiva, Uzbekistan (Fig. 1.1). The progress in numerical analyses continued until today and will continue further. The work of Brezinski and Wuytack [2] depicts the historical development of numerical analyses that gives an idea about where the roots of the subject multiphysics modelling has been established. Accordingly, it was John von Neumann (1903–1957) [3] who recognised for the first time the power of scientific computing using computers in solving complex problems, which is required to handle the complex interactions.

Hence, computer aided modelling has become more and more popular. Today it has been an essential tool for predicting and simulating

**Figure 1.1** Script from al-Kitâb al-mukhtasar fî hisâb al-jabr wa l-muqâbala and the modern statue of al-Khwarizmi at Khiva, in Ouzbekistan. Photo Alain Juhel [4].

the multiphysical behaviour of complex engineering systems. Parallel to the high-performance computer hardware development, there have been many software on the market, comprising various numerical algorithms to solve such complex problems.

These advances have contributed to an upsurge of multiphysics modelling, which has been increased since coupled phenomena such as thermal, flow and mechanics very often occur. This makes multiphysics modelling a vital component in the design development and optimisation of materials, components and systems. Today, most of the commercial programs have gained common acceptance among engineers in the industry and scientists. These extend increasingly their capabilities to take into account the interactions of various phenomena.

This will avoid the extreme simplifications on problems that are usually performed. Because the devised solutions to complex multiphysics problems via simplified mathematical models or ignored physical complexities will have an effect on the results. Moreover, to perform and evaluate the predictions still requires expertise from interdisciplinary fields comprising computer science, mathematics and many engineering fields. This is comparable to someone who knows how to read and write and uses any means of tool, such a software, pencil, typewriter to convert his knowledge into practice. Hence, the synergy between theory and the practice of multiphysics is extremely important.

But why should we consider using multiphysics modelling? First of all, there are several technical advantages employing multiphysics modelling. It is capable of reducing the time and costs of the pre-development phase of a product, substantially. Using experimental measurements are in most scientific and engineering cases very prohibitive. Setting up and performing those tests may take very long operation times. For instance, the ability to control and investigate large-scale systems or very complex systems is very difficult. To perform measurements or safety studies that are operating under hazardous conditions and are not possible, suggests the support of multiphysics modelling and simulation. Moreover, parameter studies or long-term behaviour studies that require tremendous measurement time, cost and create large data pool may benefit from the substantial reduced efforts using multiphysics modelling. Other advantages would be the detailed visualisation and analysis capabilities given by multiphysics that will improve the knowledge about the material, component process or overall system behaviour.

In contrast, the estimate expenses for the multiphysics modelling approach are based on a suitable computer hardware (~5000 €–10000 €) and software that typically ranges upon its nature, whether it is an open source, house-in code, or a commercial one perpetual/annual licence fees. The latter one has a wide range on the market from estimated ~2000 € to 30000 €, depending on the capabilities and features required.

In overall, for most engineering design, material and system development, these costs are very good value, considering the gain, time and personnel. The most expensive investment for the organisation would be the multiphysics expert. These people are extremely qualified and have may scientific and technical skills that are gathered over many years of education, training and practical experience. It should be noted that multiphysics simulations are working in synergy with experimental scientists and should not be seen as competitive, as both complement each other. Experimental activities and results are required both for data input for the numerical predictions, as well as provide data or practical knowledge to evaluate the accuracy of the analyses.

## 1.1 Multiphysics Classification

Multiphysics modelling of components materials and systems can consider interactions of a single discipline such as fluid flow coupled with heat transfer, chemically reacting flow etc. However, it can also comprise physics from various disciplines that range from solid mechanics to electromagnetism. The procedure for a coupled analysis depends on which fields are being coupled, but basically two main methods can be used namely, the load transfer and direct method.

The direct method usually deals with just one analysis that uses a coupled-field element type, containing all necessary degrees of freedom. The coupling is handled by calculating mathematical element matrices or element load vectors that contain all necessary terms.

The load transfer method involves two or more analyses, each belonging to a different field. The two fields are coupled by utilising results from one analysis as input for another analysis. Some analyses can have one-way coupling. For example, in a thermal stress problem, the temperature field introduces thermal strains in the structural field, but the structural strains generally do not affect the temperature distribution back. In such cases, there is no need to iterate between the two field solutions. More complicated cases involve two-way coupling. A piezoelectric analysis, for example, handles the interaction between the structural and electric fields together; it solves for the voltage distribution due to applied displacements, or vice versa. In a fluid-structure interaction problem, the fluid pressure causes the structure to deform, which in turn causes the fluid solution to change.

This problem requires iterations performed between the two physics fields for convergence. Moreover, the numerical grid needs to be adapted each time, as the computer domain changes its shape. The coupling between the fields can be accomplished by either direct or load transfer coupling.

Coupling across fields can be complicated because different fields may be solving for different types of analyses during a simulation. Based on how strong this interaction among the fields occurs, scientist name them such as multi-disciplinary or closely coupled. For the purpose of this book the ultimate message is that an exchange and data share is present, thus are all within the subject of multiphysics modelling.

## 1.2 Multiphysics Methodology

The methodology of executing any kind of multiphysics analysis encompasses in general, a sequence of processes. These can be grouped in three main sections. The pre-processing, the actual

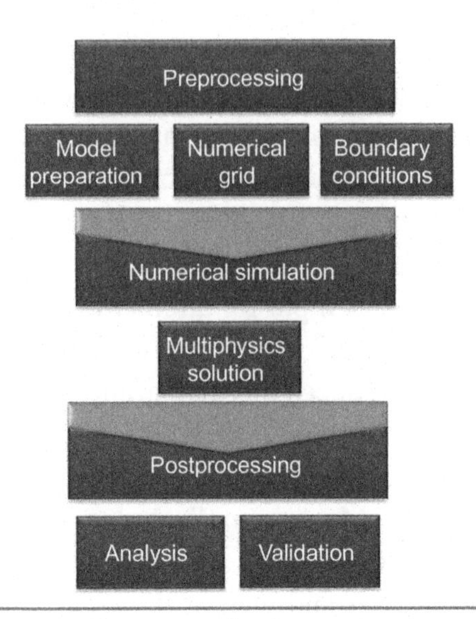


**Figure 1.2** Steps to perform a multiphysics analysis.

multiphysics solution and the post-processing section. Each section needs to be performed to end-up in a numerical model that can be utilised further for various investigations. Fig. 1.2 demonstrated the usual procedure of performing a multiphysics analysis.

The following sections will briefly examine the function and procedure of each of these core elements.

### 1.2.1 Pre-Processing

The pre-processing section is the introduction of the physical problem into the computational domain. It is this section where a digital representation of the idealised geometry will be either created, edited or readily imported from third party computer aided design programmes.

The domains are defined according their types such as solid or fluid bodies. The fluid body can be imagined analogue to a pudding mould. The mould is the solid body and the filled pudding inside the mould is the fluid body. Preparing the digital geometry for numerical analyses is indispensable. A majority of geometrical domains are generated and imported via CAD software. CAD models are subjected to some level of alteration, before they are used for numerical analyses.

This is due to the fact that even on the CAD drawings a well described volume may be present, the multiphysics software might still detect loose surfaces or several edges that are not associated with any means of the employed digital geometry. These features that lead to errors or contain useless details, need to be corrected. Because, by removing these unnecessary details, the complexity that would in anyway not contribute to the domain of interest will be reduced.

After an initial inspection, usually an operation is performed to handle these kind of issues. This procedure is known as *clean-up* in multiphysics modelling. Some analysts consider to reduce the overall part count, as to consider less parts of the physical system; whereas in some cases, each individual component is considerably simplified, which ultimately influences the accuracy of the analyses later on.

Multiphysics modelling has been used for improving this accuracy, thus the efforts of simulation improvement should not be underpinned by reducing the geometrical accuracy of the analysed system. Because not all kind of physics has the same geometrical requirements. Modern research and development is highly demanding and the increased demand on 3D coupled multiphysics simulation yields in accuracy and proper utilisation, representation of the physical domain. Hence, there are several important points to be considered whilst preparing the digital geometry and the cleaning-up procedure. This task should be performed in a tranquil manner, as it will influence the time and quality of the numerical grid, thus it will have a direct effect on the numerical results.

Traditionally, it is desired that the ends of the lines of a geometrical model needs to be always in perfect match, as this enables a closed shape and avoids later numerical grid problems performed in the next pre-processing task. Moreover, attention should be given that no duplicate features, including faces and volumes occur within a model. A common error an analyst faces is that tiny holes are present on model surfaces of the geometry.

These have to be removed by creating additional faces from the edge loops that define the holes, for instance. Cracks within a model are formed such that each edge in the pair serves as an additional edge for a separate face and share common end point vertices at one or both ends.

The edges are separated along their length by a small gap. These have to be removed from the model by connecting the edges that define the cracks. A "sharp angle" is defined as an edge pair that shares a common endpoint and acts as part of the boundary for an existing surface.

One of the edges in the sharp-angle edge pair acts as a common boundary between its bounded face and an adjacent face. It is this small angle formed between these edges. This kind of sharp interior angels from faces need to be removed by merging together the faces with adjacent faces. Analogue, large angles between faces can be removed by merging the faces. Hard edges also known by analysts as *dangling edges* are those present within a face but which do not constitute to any parts of the closed edge loop that circumscribes the face. Fig. 1.3 illustrates some of the errors usually present during pre-processing.

Research on generating geometrical domains is still going on, as to mitigate the errors present during the import-export of the created geometries [5,6]. This will reduce the time for cleaning-up processes and to mitigate the errors that arise when utilised in the multiphysics simulation software. Open and free sources have been developed within this field that provide the multiphysics analyst with a wide range of trial opportunities, as a troubling geometry may not show the same attitude when processed in a different tool. Table 1.1 depicts a brief comparison providing the pros and cons of sample software that have been tested by an expert CAD engineer, evaluating the performances and features of the different CAD systems that can also be of use for data transfer.

The assessment of the different criteria shows a range form of very poor (–) and poor (-) about average (O) till good (+) and very good fulfilled (+ +).various opportunities can be of benefit. Of

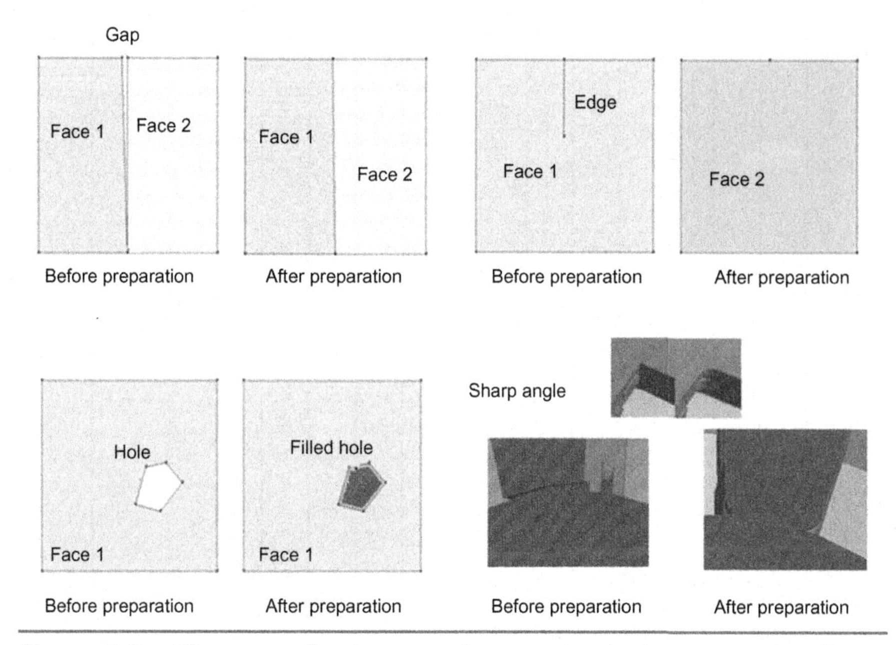


**Figure 1.3** Clean-up of a generated geometry during pre-processing.

course it only gives the reader an insight on current cost-effective opportunities that might be of benefit to consider, as well as point the functions some analysts give attention to.

The second important step in pre-processing is again a crucial one, affecting directly the overall size of your simulation model, as well as the solution time; hence, the quality of the predictions are greatly affected. This step is known as *numerical grid generation* or *mesh generation*. The ultimate goal of undergoing this step is to partition the considered physical space into small sized geometrical sections named as cells, elements or zones; over which, conservation laws can be applied and equations, describing the physical processes can be solved on.

In this procedure, it has been targeted to approximate and fit the geometrical domain within a polygonal or polyhedral frame. Depending on the discretisation method- i.e. the numerical solution method, these geometrical features have a tetrahedral, hexahedral, pyramid, or an arbitrary polyhedral form that have been used in various applications and are still progressing [8–23].

**TABLE 1.1** Performance Comparison of Some Cost Effective CAD Tool Opportunities for Multiphysics Analysts [7]

| CAD-System | Weighting factor | Autodesk 123D *Version 1.6* | FreeCAD *Release 0.15* | Sketch UP *Version 2015* | Onshape *Release 1.35* | Blender *Release 2.74* | Commercial CAD-System |
|---|---|---|---|---|---|---|---|
| **Ease of Use** | | | | | | | |
| Usability | 3 | O | – | + | ++ | – | + |
| Tutorials | 2 | ++ | + | ++ | + | + | ++ |
| GUI | 1 | + | – | + | + | – | + |
| **Scope of Functions** | | | | | | | |
| Dimensioning | 3 | O | + | – | ++ | + | ++ |
| Hole/Thread | 1 | O | + | – | + | O | ++ |
| Standard parts | 1 | – | – | – | – | – | ++ |
| Textures | 1 | – | – | ++ | – | O | + |
| Interfaces | 2 | – | ++ | + | ++ | + | + |
| Assembly | 1 | + | + | + | ++ | + | ++ |
| 2D-Drawing | 1 | – | ++ | + | ++ | – | ++ |
| **Professional Version** | | | | | | | |
| Availability (Yes/No) | –/– | N | N | Y | Y | N | –/– |
| Licensing (payment/fee) | | | | P | F | | |

The well known finite volume method (FVM) [24–27] can utilise any of those geometrical shapes, whereas the finite element method (FEM) [28,11,29–31] is traditionally not utilising polyhedral forms, for example. In such a generated mesh, each geometrical shape that contributes to the overall domain approximation contains points. These points containing data have a fixed number of neighbours. The communication between those neighbours is used to define mathematical operators. Hence, using the operators, it is possible to build equations and solve the problem. The efforts are worthy, as this section may in some complex geometries consume more than a half of the total simulation time. This depends on the expertise and practice of the analyst.

In the traditional mesh generation procedure, an important issue has been, what kind of element or cell to use, as well as how to evaluate its performance. First of all, the choice of the element will influence the element count, solution convergence and accuracy, as well as the overall simulation time. The type of mathematical problem such as curved, parabolic etc., as well as the particular physical application will also have an effect on which type of element or cell the analyst should stick on, which is far beyond the focus of the book and readers should refer to advanced research articles on those topics as well as perform *mesh independence tests*, using available element, cell types.

However, to give the reader an idea, how different the performance of element types may deviate from each other for a certain application with same boundary and operating conditions, Fig. 1.4 depicts a sample comparison, regarding element count, convergence and simulation time as well as memory usage.

The comparison illustrated utilises a polyhedral Fig. 1.4 (a) and a hexahedral cell-element type depicted in Fig. 1.4 (b). The sample velocity contour lines show that the iterations performed using the polyhedral grid still requires some more for convergence, which is supported by the convergence residual lines that drop without noise but still not having reached a steady state. The hexahedral mesh calculated the solution in shorter time, which for both analyses has

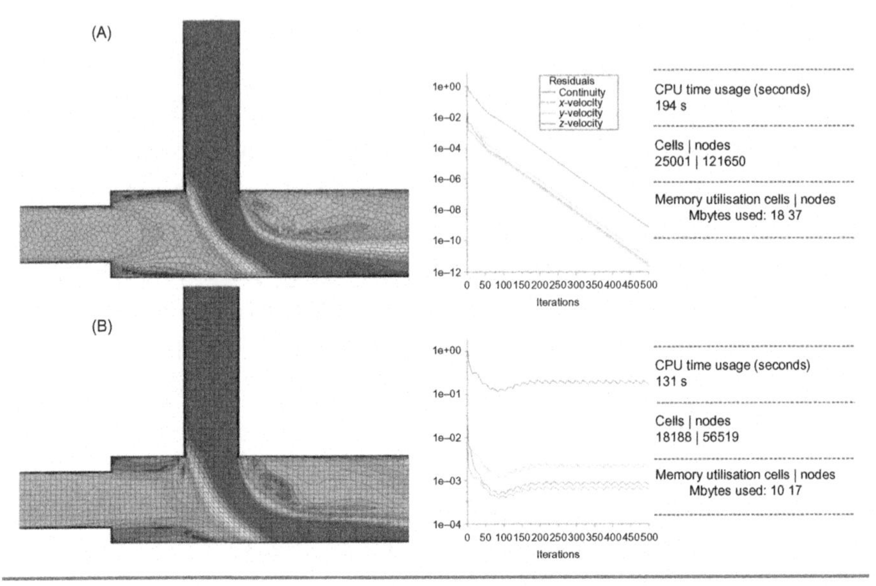


**Figure 1.4** Performance analysis employing two different element types.

been reasonable fast. The solution reached also convergence within the iteration number.

The utilised cell and node count were less compared to the polyhedral form. Also the required memory has been less for the hexahedral case. It should be noticed that these are not general statements. Especially, in complicated large scale 3D geometries that include complex physics, these data may differ. The message to the reader is that the analyst should be aware that different elements will affect their solutions, thus they should evaluate the predictions carefully and interpret the differences in results, computational performance and time, memory and the ultimate grid size prior processing for the final use of the model for further analyses.

In some kind of multiphysics applications such as fluid structure interaction (FSI), flow induced vibration, or in situations for instance, where materials can crack or large deformations may occur, the spatial domain requires to change, deform and the connectivity of the grid point- *nodes* can be difficult to be maintained without introducing errors into the simulation results. Due to this deterioration and movement of the mesh, the mathematical operators that are defined on it may no longer give accurate values. As movement of the domain occurs, elements/cells can become skewed, which can lead to numerical instabilities. In these kind of analyses, it is becoming difficult to obtain a reliable solution, using the conventional meshing procedure and special handling is required.

To mitigate these issues, either meshing the domain new i.e. *re-meshing*, which is expensive and still may encounter errors is utilised, or the computational domain has been considered, as being able to dynamically move. Several research groups have been developing numerical algorithms to handle these kind of issues [32–34]. In general, either an analogy from physical processes like for example elasticity problems are adopted and the computational domain has been considered as a deformable body [35], or interpolation methods are used. The former one is the basic one that still can encounter some errors; since all the data on the grid points must be *mapped* onto a new and different set of data points.

The second method considers the movement of the interior nodes for interpolation from the displacement of the boundary nodes. In this approach, the relocation of grid points is carried out on a node by node basis. Hence, no knowledge of mesh connectivity is required. Schemes, utilising both approaches are also presented in the literature but are beyond the scope [32]. The so called *mesh-free* or *meshless* methods are particularly useful, where the complexity of the nonlinear geometry or material behaviour makes the generation of a numerical grid prohibitive and the solution very difficult.

These issues motivated the research to mitigate the constraint of grid connectivity. Actually, the initial attempts started already in the 1970s with methods like the smoothed-particle hydrodynamics (SPH), which can be considered as the roots of the modern adaptive mesh refinement methods. But even these refinement methods are still based on an initially fixed grid spacing, which is during the solution re-fined at certain sub regions, according to the requirements of the problem.

Over the decades, the progress in completely mesh-free or the integration of some aspects of mesh-free methods has continued and with the improvement of the computer technologies will still be pursued. Discussions within specific types of algorithms and methods is left to classical texts and articles, comprising numerical methods, as it will be beyond the content of the book [35–42]. But the reader should be aware of its importance.

Most of the developments are case sensitive and it is hard to say, which methodology delivers the best solution. In practice, the analyst considers the time, cost and accuracy specifications of the project. Thus, the validation using any means of experimental data provided for the particular application is the best indication to evaluate the employed model. Even though, providing experimental data in highly complex processes is difficult, the multiphysics analyst should focus on learning and to comprehend the conventional meshing methodology as to build up a strong fundamental understanding. The practice and experience supports the analyst to build an instinct to improve the numerical grid.

Having elucidated some important aspects of the meshing procedure, the following section will focus on another important feature that is of great practical importance. Particularly, scientists, engineers or students having not reached yet a certain expertise level, usually have difficulties to evaluate the validity of the produced numerical grid. It has been targeted to shed light on this topic, including practical examples.

In general, a *high-quality mesh* is very important, in order to minimize the errors in the solvers that lead to numerical diffusion, incorrect predictions and inaccurate results. Distinguishing high-quality from bad-quality is a blurred determination of whether the mesh will properly utilised during the simulation or not. What matters actually is how accurately the multiphysics predictions reflect the reality. There are many parameters that need to be considered. These include but are not limited to the solver's numerical algorithm and the physics to be computed. But the analyst requires a judgment far beyond an intuitive evaluation. Thus, it has been decided to set-up three main categories that will help to identify and interpret the level of *validity* of the generated mesh. Accordingly, the evaluation procedure should account for the following criteria:

1. Resolution
2. Distribution
3. Mesh quality

It should be noted that like within a multiphysics solution stage, there are interactions among these criteria; thus, the parameters that affect the mesh evaluation, interact and influence each other. Therefore, the analyst requires great attention to end up with an optimum operation. This is another indication why the analyst is the most expensive investment.

To interpret the first two criteria, several parameters are taken into account. These depend on the overall meshing process i.e., used meshing methods, technical features like number of elements, advanced size functions, local refinement, time step etc. and the user meshing strategies to conduct a specific type of multiphysics analysis. Moreover, the used mesh should ideally capture the defined geometry (imitating the physical domain of interest) with sufficient detail. To practically pursue with the problem and to be able to evaluate the resolution of the mesh, a particular fixed value for each directional axis is chosen and added to the mesh that determines its resolution i.e. actual element number. This assigns a certain number of elements along each coordinate axis. Then, the so called convergence control can be performed for a chosen variable.

A first analysis of a parameter like velocity, stress or deformation etc. (depending on the physics) can be plotted. To simplify the understanding, imagine such a simple problem such as it was examined in Fig. 1.4. Imagine the pure hexahedron elements have been used. After performing the initial analysis, the same analysis and numbers of the fixed element count should be created using for example tetrahedron elements. One can observe the effect of the element basis functions such as linear, quadratic etc., as well. These will give an indication about your performance of the elements.

Furthermore, once you increase the multiphysics mesh resolution step by step by increasing the element numbers makes the results of your chosen variable converge towards the true solution (true compared to what? The validation process is the answer to this).

As soon as the results between two refinement steps don't change by more than some arbitrarily chosen threshold (depending on the application say a deviation of e. g. 2 %), we accept the mesh resolution as sufficient to be used in further analyses. This strategy will utilise the computational resources most. The mesh distribution is another feature the analyst needs to consider. The generated mesh requires certain angles between the normals for adjacent mesh elements. If the angle and the associated shapes of the geometrical feature differ from a specified angle like for example a $60^\circ$ in an equilateral triangle, the system of equations become stiffer slowing the convergence of the solution.

Furthermore, how many cells/elements are considered in small gaps between geometric entities, as well as the step size change or growth rate that specifies the increase in element edge length the neighbouring element should have in the next layer from the edge or face affect the validity of the generated grid. Several algorithms are developed to control these kind of parameters; however these range from software to software.

Those criteria so far require actually expertise in inspecting the grid qualitatively, visually; whereas the last criterion is more based on a quantitative evaluation, providing the analyst with a good initial starting point for the mesh quality validation. The so called metrics tools can quantify the mesh validity using algorithms of various type [43–49]. Different metric criterion like aspect ratio, orthogonality, skewness, least squares gradient, accounting for the orientation and proximity of neighbour cell centroids, smoothness etc are widely implemented in simulation software.

Different software and analysis types set different quality methods and criteria. It should be noted that good metrics doesn't mean necessarily a high-quality mesh. The following sample, considering a fluid flow problem utilised in Fig. 1.4 will shed light on such an effect.

Thereby the reader should notice the multiple mesh investigation steps within a certain problem solution.

The same fluid flow problem with two air inlets, releasing air from top and left into a domain accompanied by one outflow has been considered. A symmetrical half section of the 3D problem has been used for the CFD solution, considering air flow. Three different numerical grids have been utilised, each using the quantitative skewness metrics of 0.75 respectively. Thereby, the number 1 should indicate a bad mesh quality and the lower values the good quality. All the results have been scaled using the same legend and in gray, thus a comparison can be performed easily, by visualising the result on the symmetry plane section. Fig. 1.5 illustrates the investigated cases.

The results demonstrate the numerical grid and the superimposed contour plot lines of the fluid flow velocity distribution. Each simulation result is accompanied with its residual convergence lines that normally gives an indication about the convergence of the solved equations of continuity and momentum. All three analyses have been performed for an iteration period of 500. The differences are clearly indicated. The first analysis (Fig. 1.5 (a)), utilising a coarser resolved grid indicates a very good convergence behaviour compared to the other two residual curves depicted in Fig. 1.5 (b) and Fig. 1.5 (c) that are comprising noise: however, when the plots are carefully compared, it is clearly visible that the best resolved grid illustrated in Fig. 1.5 (c) captures the flow distribution with the best details. The same metrics of 0.75 skewness resulted in different degree of resolved flow areas. In the first two analyses, capturing the flow development and the details of the inlet region and above the elbow flow profile is misleading.

The authors experience shows that the metrics alone and its associated convergence residuals are not sufficient for a quality validation and the resolution-mesh density has a more important impact on the overall capturing of the flow distribution. A numerical grid with proper resolution has shown in most cases to overcome the convergence issues and yield to more accurate results.

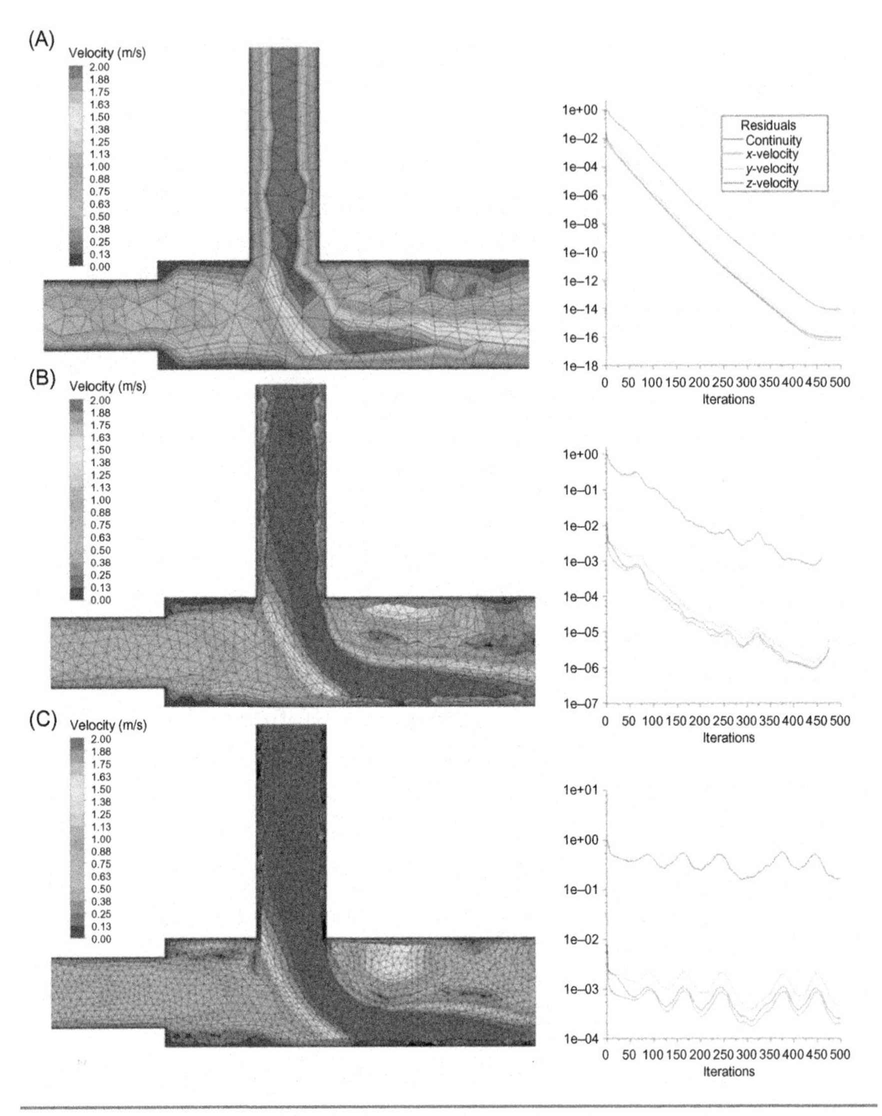


**Figure 1.5** Demonstration of the mesh quality effect on the flow distribution.

Once the prohibitive numerical grid has been created, certain conditions have to be determined as a next step. The physical problem needs to be comprehended by the digital representation i.e. employed computer model. These are the boundary conditions, constraints, operating conditions that describe the frame of the problem, where input data for the numerical simulations are introduced. The boundary conditions vary according the employed physics, as for fluid dynamics, for instance, boundary conditions such as velocity inlet, pressure outlet and volumetric flow rates define how a fluid enters or leaves the computational domain; whereas heat flux or film coefficients determine the interchange of energy between the computational model and the environment such as air, water etc. that is subject of heat transfer.

Hence, the boundary conditions link the multiphysics model with its surroundings. Without them, the simulation is not ready for solving the problem, as the governing equations cannot be solved successfully. Another important feature of the boundary conditions is that they can be applied in either steady state or transient states.

The steady case is of great importance in engineering applications, where a certain behaviour of the system will be retained, as time passes. It considers that variables have reached a certain level and are not changing over time. When a particular behaviour is changing over time, a steady state is no more valid and the so called transient case takes place. Engineering processes like heating-up, cooling or cyclic loading scenarios are typical examples where time plays a major role. This kind of analyses require an additional kind of input condition that describes the initial situation of the problem prior proceeding over time. Defining this situation is named as the initial conditions. Unlike boundary conditions, initial conditions are only utilised at the beginning of the analysis. They are primarily used for transient analyses, but are sometimes employed for steady state analyses too. Due to the large application area of multiphysics, obviously there are various initial and boundary condition opportunities to be applied. However, upon their application one could group them mainly either into conditions suitable for fluid zones or solid regions. Some type of boundary conditions such as pressure are suitable for both applications. At this point, instead of going into details of several boundary condition types, it is believed that shedding light on some aspects that may require special care whilst applying on the multiphysics model, would be more beneficial to the reader.

If we consider multiphysics problems comprising fluid domains, the flow boundary conditions characteristically represent a quantity or state at the model opening. These are applied in 3D models on surfaces, whereas in 2D representations, the edges of the models are utilised for the successful boundary condition application. The important aspect is that the analyst needs to think about which real-life loading scenario is tried to be depicted in the model. One can employ the boundary conditions to an infinitesimal small area or on single nodes, as well; however, these kind of concentrated conditions would result in high gradients, which is the reason why distributed loads such as on edges or surfaces are employed and actually are more realistically imitating the real physical boundary conditions.

The following structural analysis example emphasizes the importance of the correct applications of boundary conditions on the model geometry. A quadratic sample plate of 1 m x 1 m x 0.1 m dimensions is subjected on one side to a force of 20.000 N and is constrained on the other right surface. In this kind of scenario, it is important to interpret whether or not the used loading area is reasonable and reflects the real-life case. Fig. 1.6 (a) illustrates the first case in which the applied force is utilised as concentrated load i.e. the force is applied to grid nodes. The demonstrated total deformation results show that the plate corners are subjected to higher deformation. However, this is due to the applied boundary effects. The applied forces on the corner nodes will act only on half of the element edge, thus results are affected. This is the reason why local maximum values are observed in the vicinity of the corners.

Fig. 1.6 (b) depicts the distributed loading case to mitigate the previously observed concentrated maxima points. In this case, the load has been applied as a surface load on the whole surface. (As the depth is very thin the reader can interpret the case as if it was in two dimensional, thus can imagine the edge has been chosen as line load. The results demonstrate now an accurate representation of the case with the corrected initial distribution on the side where the force has been applied.

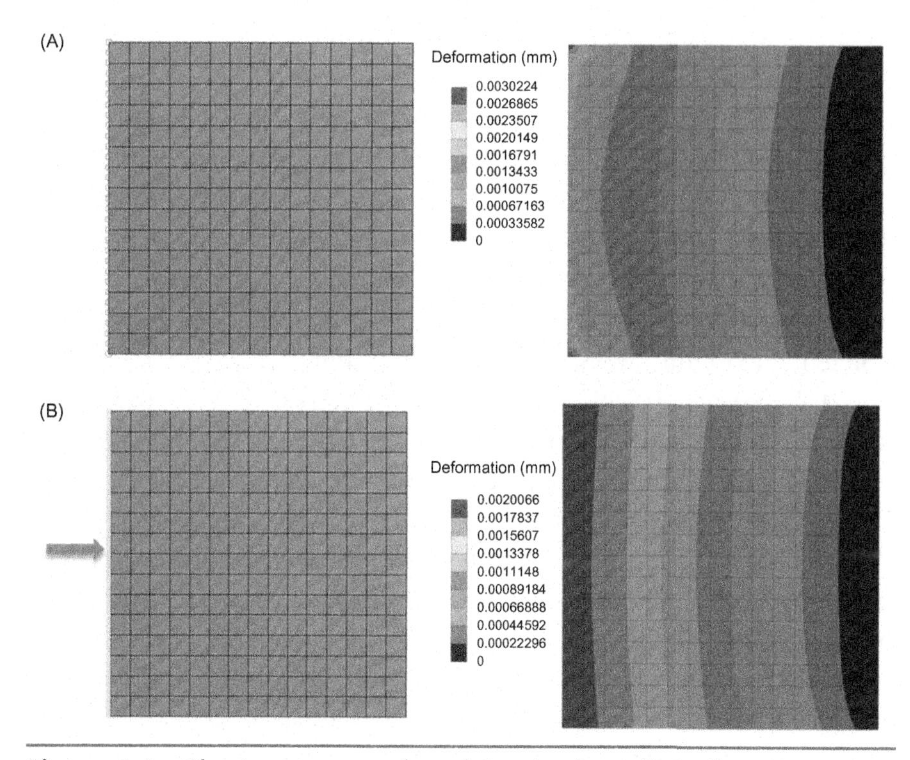


**Figure 1.6** The importance of applying loads realistically: (A) concentrated loading and (B) distributed loading.

One important feature in applying boundary conditions is the use of constraints. Constraints or the so called supports are used to restrain structures against relative rigid body movements. In which location and of how many degrees of freedom one should employ the constraints are usually of practical concern. This is particularly in regions with solid regions of great interest. The reason is, when the model is not sufficiently restrained, once it is exposed to loads, it will allow infinite displacements. Thus, rigid body motion will occur and ultimately the solution will stop, giving errors.

To mitigate this problem, based on the employed 2D or 3D approach, the model needs to be restrained. This requires that for a 2D approach, the model must be supported against two translational movements i.e. along x and y, and one rotational movement around z. Therefore, the minimum number of constraints that has to be applied in two dimensions is three. The minimal number of restrictions for the three dimensional case is six.

In practice, problems related to structural mechanics require particularly the careful use of constraints. The stiffness of the model will be influenced by the constraint type and location. Fig. 1.7 depicts an example of the effect of constraining the same component differently. A force of 5000 N is applied on three steel samples each constrained on different areas. The Von Mises Stress results depicted maximum stresses of 0.25 MPa (Fig. 1.7 (a)), 0.011 MPa (Fig. 1.7 (b)) and 0.049 MPa (Fig. 1.7 (c)), respectively predicted using the same material properties. The exaggerated scaling for the predicted Von Mises Stress contour plots are demonstrated to show the movement of the bodies.

As the maximum value differences are large, they are not scaled and plotted with the same legend. Thus an optical colour comparison, regarding the stress level would be irritating the reader. Aim has been to emphasise that despite the same magnitude of force has been used in all cases, selecting different areas to apply the constraints obviously effects the vicinity where the maximum stresses arise. Moreover, it should turn the attention of the reader on carefully using the constraints during modelling.

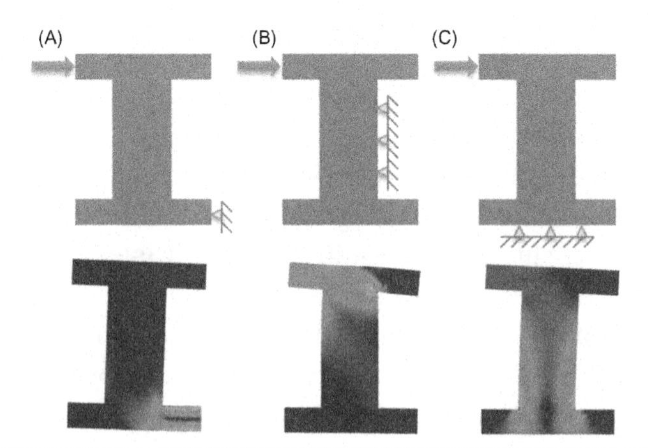


**Figure 1.7** The effect of using the same magnitude of force with different constraints: The maximum level of stress results, as well as the distribution will be affected.

Another important issue requiring special care is the application of symmetry boundary conditions. Applying a quarter, half or a portion of the physical model reduces the degree of freedom, as well as the solution time. However, the geometrical simplification, whilst considering symmetry conditions is limited to the digital representation of the computational model.

In order that the software comprehends that the employed model represents just a portion of the whole model, these conditions have to be delegated by the analyst. Any out of symmetry plane degree of freedom must be constrained as to avoid the node motion and applied to the model.

Once introduced to the model, the grid nodes at the symmetry planes are not allowed to move or rotate. This is how the symmetry boundary condition is working; however, special care is needed regarding the use and limitations of the symmetry approach. This is particularly the case, where some heat radiation models or dynamics analyses like modal analyses are employed. Some modes such as out of phase modes might not be predicted

On the other side, care should also be taken when using specific boundary conditions types. the analyst needs to understand, if the assumption of the analyst using the certain boundary condition type matches with what the software is assuming and executing. To shed light on this issue, a typical example using the boundary conditions "mass flow rate" employed on a 2D problem is given. The given rectangular structure is assumed to focus on the problem rather than geometric complexity. When a given mass flow rate is introduced to the 2D problem, as previously mentioned the condition is applied on the edge of the domain filled with say for example air.

The importance to comprehend how the software handles this condition is here clear. The analyst needs to think that the problem is 2D and an edge is used. The definition of mass flow rate is that it is related to velocity through the continuity equation. But how can this be the case? The two terms are related employing the area of the inlet region. This is exact the important part. How will the velocity be calculated using the surface area if the defined problem is just an edge and not a surface! Some software utilise for this purpose a default width of 1 m! If the user is not aware of this, the problem will consider a calculated surface area using 1 m width. Therefore, the actual velocity calculated will be drastically different compared to the real velocity.

The solution for this is that either the user should correct the used default value of 1 m, if this is possible. Another way might be to employ velocity boundary conditions. Moreover, the analyst can also extrude the 2D geometry through a small depth into a 3D model. This would enable the use of a closed face at the inlet and calculate a correct velocity. As the third dimension would be such thin that the fluctuation of the variables in the third dimensions would be negligible small, hence predicting the profiles such as in a 2D state.

Another useful inlet boundary condition is the utilisation of a volume flow rate. It is most often useful if the density is constant throughout the numerical solution.

### 1.2.2 Numerical Simulation

The numerical solution part is where the software is proceeding with the actual multiphysics calculation and solve the completely defined model. It is important to comprehend the solution procedure, as to have control on the predictions. There has been a vast amount of literature, comprising various aspects and kinds of numerical methods, thus the Chapter will not give particular focus to the detailed solution techniques, but will elucidate some points that require attention. Moreover, useful references on general aspects [50–63], as well as on specific numerical applications [64–74] are given. Within the different solution methods, the highly sought after analytical methods are unique to certain problems and too difficult to be employed in complex physics. The heuristic solutions that are mostly based on trial-error approaches mitigate their utilisation in problems such as complex multiphysics analyses, as well. This is the reason why the solution method chosen for multiphysics problems is based on numerical analysis.

In particular, the governing equations used in multiphysics are highly non-linear and the utilised boundary conditions are usually defined as functions, which forces the solution process to be iterative (Jacobi [75], Gaussian [76], Galerkin [77] etc.). This is regardless of whether the problem is time dependent or not. The iterative procedures give approximations of the exact problem solution, thus are more favoured compared to the direct solutions that enable just partially exact solutions. Of course, the analyst should carefully think about the suitability of the chosen solution method.

Concerning, if the assumptions and simplifications of the problem are transferrable to the solution method and if expert knowledge about the use and application of the method is required within the software to perform settings, changes i.e. the parameters the solution method depends on. Actually, one of the major advantages of using numerical methods is that there is virtually no limit to handle the complexity of problems that can be solved and this is an important feature in multiphysics analyses. However, to numerically solve the problem, an initial solution i.e. a guessed solution is required at the beginning phase of the solution procedure. Thereafter, the numerical equations are used to produce a more accurate approximation to the numerically correct solution, which is one in that all the variables satisfy the governing equations.

The new approximation, the so called updated solution is then used as the new starting solution and the process continuous in a repeating way, until the error in the solution is sufficiently reduced. Each time the process is repeated, it is said to have accomplished an iteration. And the measure of errors are known as residual errors or residuals, which need to be controlled. This requires knowledge and practical expertise by the user to comprehend and take control over the solution process. By computing the solution, the solver requires sufficient iterations to achieve *convergence.*

Practically, the convergence could be described as "*the property of a numerical method to produce a solution, which approaches the exact solution as the grid spacing, control volume size or element size is reduced to zero"* [68]. All discrete conservation equations need to be obeyed in all the grid cells/elements to a specified tolerance i.e. residuals. This measures and evaluates the imbalance of the calculated solution. Balances for all the overall governing equations and scalars are achieved.

The practical requirements of the numerical solution part for the analyst is definitely to gather hints to take control over the convergence, which is linked to the solution. Therefore, some practical aspects of this problem is covered to shed light on the issues most scientists and engineers face, whilst handling multiphysics analyses. Thereby, it should be noted that the ultimate goal for the analyst is to keep the calculations robust, fast and efficient. However, the multiphysics solution techniques are problem specific and cannot be generalised.

The effect of the utilised algorithms play here a major role. From a mathematical point of view trying to investigate the convergence properties of solutions using deterministic or stochastic iterative methods, for instance, will give an insight and improve the understanding of the analyst about the theory; however, it will not solve the numerical problems the analyst faces using a certain software in practice. Hence, some practical guidelines are required that may support the multiphysics analyst.

The analyst usually reads general statements like a decrease in residuals of a variable by "some" orders of magnitude would lead to a sign of convergence. These are vague definitions that are not valid for all types of problems, nor are they same for all software. Therefore, the user needs to employ a different approach that enables more control and monitoring. Most of the software have integrated a utility that allows one to monitor the fall of residual curves during the analysis.

The convergence criterion are in that case set by default values like $10^{-6}$ for variable solutions, so the residual fall below the given tolerance at each time step, step-size can be monitored. However, relying solely on this method is not sufficient.

Of course, the most accurate and valid evaluation would be the validation/verification process performed during the post-processing section. In practice, it is common to use simplified analytical solutions for comparison. However, during the solution the analyst may follow some rules.

The idea is to reach a stable and consistent solution result that does not change anymore, which is an indication of convergence. This is particularly the case when the solution is linear, whereas in unsteady cases, this change in the variables should be investigated by reducing the time and distance steps. Hence, all conservation equations involved in the problem need to satisfy this stable situation. One of the best practical monitoring techniques is to define a point or surface within the computational domain. The time or iteration history of variables, as well as the residuals for the conservation equations are then monitored for the pre-defined location. Another hint would be to compute an initial solution, using a simpler first-order discretisation scheme and to run the simulation.

For the final prediction, higher-order upwind discretisation schemes can then be followed. Also, the values used for solution initialisation should be reasonable. Zero values are always problematic in numerical methods; thus, in species transport for example, where values for species are zero, a very low value like 0.01 for initialisation should be given as a starting point, as to mitigate an immediate divergence of the solution. If the monitored residual lines indicate that the solution is converged and finished, but the analyst observes that the solution is still changing or for example shows imbalances, this clearly indicates the solution is not yet converged. Solving this kind of issues are best learned from experience.

Some software include terms called relaxation factors that assist the solution in convergence. The lowering and use of these terms require great care such as starting from an initial solution and gradually increasing etc. The solution instabilities can also arise with an ill-posed problem, thus it is always wise to go through the employed physical models and boundary conditions, as well. The importance and impact of the numerical grid specifications should here be once more strongly emphasized. Ensure that the problem solution is grid-independent.

Cells or elements which have large aspect ratio should be improved (certain criteria and limits utilised within used software should be understood). As the numerical solution errors are associated with computing the cell gradients and cell face interpolations, it is helpful to decrease cell size and cell size variations. These will minimize the interpolation and truncation errors. The chosen boundary locations; especially, the choice of the domains of inlet and outlet regions, distance of constraints to the applied force region etc. will significantly affect the results; therefore, require careful observation.

Ultimately it should be noted that all attempts to reach a robust and converged solution is not necessarily resulting in correct predictions. It is always wise to inspect and evaluate the solution by using available data, physical principles i.e. give attention to the validation/verification procedure.

### 1.2.3 Post-Processing

The last constituents of the multiphysics modelling methodology is the post-processing section that covers briefly, the demonstration and evaluation of the predictions, compare and validate the computed results with experimental measurements or verify using suitable mathematical tools. Moreover, decisions on the potential use of the multiphysics model in further studies or if optimisation of the model is required are made within this section. Due to the advancements in computer hardware and improvements on computing efficiency, the calculation time for large scale multiphysics problems have been significantly reduced. However, the created simulation data management, as well as the exploration of these results still require care within the post-processing section. Therefore, it is wise to give some insight on this topic, prior getting into important aspects within the post-processing section.

Today, most of the simulation life cycle management or data management tools are mainly used to organise and support the user to extract data for their need. Various software companies are integrating intelligent data compression techniques to store and process data efficiently. However, these data need to be explored carefully. It is important to detect and interpret valuable correlations among variables, interpret the information the results give, first of all. This is the reason why the evaluation part of the whole data is important; otherwise, it would be similar to millions of raw data created during experimental measurements. It is exceedingly important to take advantage of these capabilities in simulation, as to save time and costs.

This is essential to have an efficient workflow and to proceed with choosing the best way to demonstrate the results and to reach the ultimate goal of the performed analyses. If the analyst fails to effectively use these capabilities, this will encourage people to blame the simulation technology as being colourful pictures and prevents to use the full potential of the simulation technology.

## References

[1] Rosen F. (trs. Muhammad ibn Musa Al-Khwarizmi: Algebra. n.d.

[2] Brezinski C, Wuytack L. (Luc) Numerical analysis: historical developments in the 20th century. Elsevier; 2001.

[3] Aspray W. John von Neumann and the origins of modern computing. Cambridge, MA: MIT Press; 1990.

[4] Modern statue of al-Khwarizmi at Khiva, in Ouzbekistan. Photo Alain Juhel n.d. www.muslimheritage.com/article/contribution-al-khwarizmi-mathematics-and-geography.

[5] Chow P, Kubota T, Georgescu S. Automatic Detection of Geometric Features in CAD models by Characteristics. Comput Aided Des Appl 2015;12:784–93. Available from: https://doi.org/10.1080/16864360.2015.1033345.

[6] Gindis E, Gindis E. Chapter 16 – Importing and Exporting Data. Up Run. with AutoCAD 2017, 2017, p. 409–22. Available from: https://doi.org/10.1016/B978-0-12-811058-4.00016-5.

[7] Junk S, Kuen C. Review of Open Source and Freeware CAD Systems for Use with 3D-Printing. Procedia CIRP 2016;50:430–5. Available from: https://doi.org/10.1016/j.procir.2016.04.174.

[8] Sohn D, Jin S. Polyhedral elements with strain smoothing for coupling hexahedral meshes at arbitrary nonmatching interfaces. Comput Methods Appl Mech Eng 2015;293:92–113. Available from: https://doi.org/10.1016/j.cma.2015.04.007.

[9] Sauer G. Diagonal mass and associated stiffness matrices for isoparametric 5-node quadrilateral and 9-node hexahedron elements. Finite Elem Anal Des 1993;13:37–47. Available from: https://doi.org/10.1016/0168-874X(93)90005-B.

[10] Rjasanow S, Weißer S. FEM with Trefftz trial functions on polyhedral elements. J Comput Appl Math 2014;263:202–17. Available from: https://doi.org/10.1016/j.cam.2013.12.023.

[11] Remacle J-F, Gandham R, Warburton T. GPU accelerated spectral finite elements on all-hex meshes. J Comput Phys 2016;324:246–57. Available from: https://doi.org/10.1016/j.jcp.2016.08.005.

[12] Ales Z, Knippel A, Pauchet A. Polyhedral combinatorics of the K-partitioning problem with representative variables. Discret Appl Math 2016;211:1–14. Available from: https://doi.org/10.1016/j.dam.2016.04.002.

[13] Cantin P, Bonelle J, Burman E, Ern A. A vertex-based scheme on polyhedral meshes for advection–reaction equations with sub-mesh stabilization. Comput Math with Appl 2016;72:2057–71. Available from: https://doi.org/10.1016/j.camwa.2016.07.038.

[14] Castillo JMF, Papini PL. On isomorphically polyhedral $L\infty$-spaces. J Funct Anal 2016;270:2336–42. Available from: https://doi.org/10.1016/j.jfa.2016.01.003.

[15] Gao L., Zhang Q., Zhu M., Zhang X., Sui G., Yang X. Polyhedral oligomeric silsesquioxane modified carbon nanotube hybrid material with a bump structure via polydopamine transition layer. vol. 183. 2016. doi:10.1016/j.matlet.2016.07.107.

[16] Huang L, Zhao G, Wang Z, Zhang X. Adaptive hexahedral mesh generation and regeneration using an improved grid-based method. Adv Eng Softw 2016;102:49–70. Available from: https://doi.org/10.1016/j.advengsoft.2016.09.004.

[17] Jaśkowiec J, Pluciński P, Stankiewicz A. Discontinuous Galerkin method with arbitrary polygonal finite elements. Finite Elem Anal Des 2016;120:1–17. Available from: https://doi.org/10.1016/j.finel.2016.06.004.

[18] Ji S, Sun S, Yan Y. Discrete Element Modeling of Rock Materials with Dilated Polyhedral Elements. Procedia Eng 2015;102:1793–802. Available from: https://doi.org/10.1016/j.proeng.2015.01.316.

[19] Jiang G. Numerical Prediction of Turbulent Flow over 3D Sinusoidal Hill Using Non-orthogonal Hexahedron Grid and AMG Method.

Procedia Eng 2012;31:57–61. Available from: https://doi.org/10.1016/j.proeng.2012.01.990.

[20] Jin S, Sohn D, Im S. Node-to-node scheme for three-dimensional contact mechanics using polyhedral type variable-node elements. Comput Methods Appl Mech Eng 2016;304:217–42. Available from: https://doi.org/10.1016/j.cma.2016.02.019.

[21] Lee SY. Polyhedral Mesh Generation and A Treatise on Concave Geometrical Edges. Procedia Eng 2015;124:174–86. Available from: https://doi.org/10.1016/j.proeng.2015.10.131.

[22] Landier S. Boolean operations on arbitrary polygonal and polyhedral meshes. Comput Des 2016. Available from: https://doi.org/10.1016/j.cad.2016.07.013.

[23] Lei N, Zheng X, Jiang J, Lin Y-Y, Gu DX. Quadrilateral and hexahedral mesh generation based on surface foliation theory. Comput Methods Appl Mech Eng 2016. Available from: https://doi.org/10.1016/j.cma.2016.09.044.

[24] Moukalled F, Mangani L, Darwish M. The Finite Volume Method in Computational Fluid Dynamics: An Advanced Introduction with OpenFOAM and Matlab. 1st ed Springer Publishing Company, Incorporated; 2015.

[25] Sonar T. Classical Finite Volume Methods. Handb. Numer. Anal. 2016. Available from: https://doi.org/10.1016/bs.hna.2016.09.005.

[26] Mazumder S, Mazumder S. Chapter 6 – The Finite Volume Method (FVM). Numer. Methods Partial Differ. Equ. 2016;277–338. Available from: https://doi.org/10.1016/B978-0-12-849894-1.00006-8.

[27] Eymard R, Gallouët T, Herbin R. Finite volume methods. Handb Numer Anal 2000;7:713–1018. Available from: https://doi.org/10.1016/S1570-8659(00)07005-8.

[28] Akin JE, Akin JE. Chapter 22 – MODEL APPLICATIONS IN 2- AND 3-D. Finite Elem. Anal. Des. 1994;461–514. Available from: https://doi.org/10.1016/B978-0-08-050647-0.50026-5.

[29] Jeong S, Lee E. Weighted norm least squares finite element method for Poisson equation in a polyhedral domain. J Comput Appl Math 2016;299:35–49. Available from: https://doi.org/10.1016/j.cam.2015.10.011.

[30] Kyosev YK. 6 – The finite element method (FEM) and its application to textile technology. Simul. Text. Technol. 2012;172–222e. Available from: https://doi.org/10.1533/9780857097088.172.

[31] Liu GR, Quek SS, Liu GR, Quek SS. Chapter 3 – Fundamentals for Finite Element Method. Finite Elem. Method 2014;43–79. Available from: https://doi.org/10.1016/B978-0-08-098356-1.00003-5.

[32] Luke E, Collins E, Blades E. A fast mesh deformation method using explicit interpolation. J Comput Phys 2012;231:586–601. Available from: https://doi.org/10.1016/j.jcp.2011.09.021.

[33] Walton S, Hassan O, Morgan K. Advances in co-volume mesh generation and mesh optimisation techniques. Comput Struct 2016. Available from: https://doi.org/10.1016/j.compstruc.2016.06.009.

[34] Pissanetzky S, Pissanetzky S. CHAPTER 8 – Connectivity and Nodal Assembly. Sparse Matrix Technol. 1984;271–87. Available from: https://doi.org/10.1016/B978-0-12-557580-5.50013-2.

[35] Stein K, Tezduyar T, Benney R. Mesh Moving Techniques for Fluid-Structure Interactions With Large Displacements. J Appl Mech 2003;70:58–63.

[36] Liu GR. Meshfree Methods: Moving Beyond the Finite Element Method. CRC Press; 2010. Available from: https://doi.org/10.1115/1.1553432.

[37] Zhang L, Liu W, Li S, Qian D, Hao S. Survey of Multi-Scale Meshfree Particle Methods. Lect Notes Comput Sci Eng 2003;26:441–58.

[38] Zhang LT, Wagner GJ, Liu WK. Modelling and simulation of fluid structure interaction by meshfree and FEM. Commun Numer Methods Eng 2003;19:615–21.

[39] Li S, Liu WK. Meshfree and particle methods and their applications. Appl Mech Rev 2002;55:1.

[40] Liu MB, Liu GR. Smoothed particle hydrodynamics (SPH): An overview and recent developments. Arch Comput Methods Eng 2010;17:25–76.

[41] Liu GR, Gu YT. An introduction to meshfree methods and their programming. Netherlands: Springer; 2005.

[42] Babuka I, Banerjee U, Osborn JE. Survey of meshless and generalized finite element methods: A unified approach. Acta Numer 2003;12:1–125.

[43] Pellenard B, Orbay G, Chen J, Sohan S, Kwok W, Tristano JR. QMCF: {QMorph} Cross Field-driven Quad-dominant Meshing Algorithm. Procedia Eng 2014;82:338–50. Available from: http://dx.doi.org/10.1016/j.proeng.2014.10.395.

[44] Loseille A. Metric-orthogonal Anisotropic Mesh Generation. Procedia Eng 2014;82:403–15. Available from: http://dx.doi.org/10.1016/j.proeng.2014.10.400.

[45] Shephard MS, Flaherty JE, Jansen KE, Li X, Luo X, Chevaugeon N, et al. Adaptive mesh generation for curved domains. Appl Numer Math 2005;52:251–71. Available from: http://dx.doi.org/10.1016/j.apnum.2004.08.040.

[46] Huang W. Metric tensors for anisotropic mesh generation. J Comput Phys 2005;204:633–65. Available from: http://dx.doi.org/10.1016/j.jcp.2004.10.024.

[47] Blazek J. Chapter 11 - Principles of Grid Generation. In: Blazek J, editor. Comput. Fluid Dyn. Princ. Appl. Third Ed. Oxford: Butterworth-Heinemann; 2015. p. 357–93. Available from: http://dx.doi.org/10.1016/B978-0-08-099995-1.00011-7.

[48] Baker TJ. Mesh generation: Art or science? Prog Aerosp Sci 2005;41:29–63. Available from: http://dx.doi.org/10.1016/j.paerosci.2005.02.002.

[49] Sirois Y, Dompierre J, Vallet M-G, Guibault F. Hybrid mesh smoothing based on Riemannian metric non-conformity minimization. Finite Elem Anal Des 2010;46:47–60. Available from: http://dx.doi.org/10.1016/j.finel.2009.06.031.

[50] Chen Z. Finite Element Methods and Their Applications. 2005. Available from: http://dx.doi.org/10.1007/3-540-28078-2.

[51] Strikwerda J.C. Finite difference schemes and partial differential equations. 2004.

[52] Liu GR, Quek SS. The Finite Element Method. Elsevier; 2014. Available from: https://doi.org/10.1016/B978-0-08-098356-1.00008-4.

[53] A.A. Mohammed. Lattice Boltzmann Method: Fundamentals and Engineering Applications with Computer Codes. 2012. Available from: http://dx.doi.org/10.2514/1.J051744.

[54] Logan DL, Veitch E, Carson C, Burrell KR, Gould V, Wagner E. A First Course in the Finite Element Method Fourth Edition 2007;147. Available from: https://doi.org/10.1016/0022-460X(91)90505-E.

[55] Munjiza A. The Combined Finite-Discrete Element Method. vol. 41. 2008. Available from: http://dx.doi.org/10.2307/20206579.

[56] Fish J, Belytschko T. A First Course in Finite Elements. John Wiley and Sons; 2007.

[57] Burden RL, Faires JD. Numerical Analysis. 2011. Available from: http://dx.doi.org/10.1017/CBO9781107415324.004.

[58] Kiusalaas J. Numerical Methods In Engineering With: Python. vol. 11. 2005. Available from: http://dx.doi.org/10.1017/CBO9780511812217.

[59] Chapra SC, Canale RP. Numerical methods for engineers, vol. 33. McGraw-Hill Science/Engineering/Math; 2015. Available from: https://doi.org/10.1016/0378-4754(91)90127-O.

[60] Peiro J, Sherwin S. Finite Difference, Finite Element and Finite Volume Methods for Partial Differential Equations. Handb Mater Model 2005; M2415–46. Available from: https://doi.org/10.1007/978-1-4020-3286-8_127.

[61] Cai Z. On the finite volume element method. Numer Math 1991;58:713–35.

[62] Barth T., Ohlberger M. Finite volume methods: foundation and analysis. vol. m. 2004. Available from: http://dx.doi.org/10.1002/0470091355.ecm010/full.

[63] Xi H, Peng G, Chou S-H. Finite-volume lattice Boltzmann method. Phys Rev E 1999;59:6202–5. Available from: https://doi.org/10.1103/PhysRevE.59.6202.

[64] Rapp BE, Rapp BE. Chapter 33 – Numerical Solutions to Transient Flow Problems. Microfluid. Model. Mech. Math. 2017;679–99. Available from: https://doi.org/10.1016/B978-1-4557-3141-1.50033-2.

[65] Chicone C, Chicone C. Chapter 16 – Numerical Methods for Computational Fluid Dynamics. An Invit. to Appl. Math. 2017;403–510. Available from: https://doi.org/10.1016/B978-0-12-804153-6.50016-6.

[66] Christensen J, Bastien C, Christensen J, Bastien C. Chapter I two – Numerical Techniques for Structural Assessment of Vehicle Architectures. Nonlinear Optim. Veh. Saf. Struct. 2016;51–105. Available from: https://doi.org/10.1016/B978-0-12-417297-5.00002-X.

[67] De Bortoli ÁL, Andreis GSL, Pereira FN, De Bortoli ÁL, Andreis GSL, Pereira FN. Chapter 6 – Numerical Methods for Reactive Flows. Model. Simul. React. Flows 2015;123–69. Available from: https://doi.org/10.1016/B978-0-12-802974-9.00006-4.

[68] Versteeg H.K., Malalasekera W. An Introduction to Computational Fluid Dynamics - The Finite Volume Method. 1995. Available from: http://dx.doi.org/10.2514/1.22547.

[69] Chai JC, Lee HS, Patankar SV. Finite Volume Method for Radiation Heat Transfer. J Thermophys HEAT Transf 1994;8:419–25. Available from: https://doi.org/10.2514/3.559.

[70] Ferziger JH, Peric M. Computational Methods for Fluid Dynamics. 2002. doi:10.1016/S0898-1221(03)90046-0.

[71] Stangl R, Leendertz C, Haschke J. Numerical Simulation of Solar Cells and Solar Cell Characterization Methods: the Open-Source on Demand Program AFORS-HET. 2010. Available from: http://dx.doi.org/10.5772/8073.

[72] Zienkiewicz OC, Taylor RL, Nithiarasu P. The Finite Element Method for Fluid Dynamics. Elsevier; 2014. Available from: https://doi.org/10.1016/B978-1-85617-635-4.00001-7.

[73] Lewis R, Nithiarasu P., Seetharamu K. Fundamentals of the finite element method for heat and fluid flow. 2004. Available from: http://dx.doi.org/10.1002/0470014164.

[74] Polycarpou AC. Introduction to the finite element method in electromagnetics. vol. 1. 2005. Available from: http://dx.doi.org/10.2200/S00019ED1V01Y200604CEM004.

[75] Falcone M, Ferretti R. Chapter 23 – Numerical Methods for Hamilton–Jacobi Type Equations. Handb. Numer. Anal. 2016;17:603–26. Available from: https://doi.org/10.1016/bs.hna.2016.09.018.

[76] Tian Z, Tian M, Liu Z, Xu T. The Jacobi and Gauss–Seidel-type iteration methods for the matrix equation AXB = C. Appl Math Comput 2017;292:63–75. Available from: https://doi.org/10.1016/j.amc.2016.07.026.

[77] Zienkiewicz OC, Taylor RL, Fox D, Zienkiewicz OC, Taylor RL, Fox D. Chapter 2 – Galerkin Method of Approximation: Irreducible and Mixed Forms. Finite Elem. Method Solid Struct. Mech. 2014;21–55. Available from: https://doi.org/10.1016/B978-1-85617-634-7.00002-8.

Chapter 2

# Multiphysics Modelling of Fluid Flow Systems

## Chapter Outline



Fluid dynamics has been one of the most important constituents of multiphysics. Even life would not been possible without the existence of fluids. Therefore understanding the behaviour of fluid flow systems has been a major field within multiphysics modelling, as well. The complex behaviour of fluids requires great care whilst mimicking in multiphysics modelling.



Multiphysics Modelling. DOI: https://doi.org/10.1016/B978-0-12-811824-5.00002-X

With all its fundamentals and different applications, fluid flow systems require special attention that has been covered in various books, journal papers and other sources, so far. For a systematic development of the current work, however, it is first aimed to give a brief overview and understanding of stand-alone physical areas, contributing to multiphysics problems. This will help the reader to keep it simple and establish a basic fundamental in modelling aspects that need attention, prior mixing and handling several topics and getting into more complex and challenging phenomena.

Hence, it is useful to start with sole fluid flow systems and classify them into typical practical occurrences. Accordingly, an insight to important modelling aspects will be given and some practical features will be discussed. It should be noted that it has been merely scratched the surface of this giant field. The target has been to introduce the important aspects from a user's point of view and to provide some guidelines that will be useful in complex multiphysics situations. The reader will improve the knowledge in handling fluidic domains, when solving multiphysics problems.

Generally, the fluid flow systems can involve a material in liquid state such as in mould or casting processes, be considered as *Newtonian* [1–4] or *non-Newtonian* [5–9], take place in *particulate solutions* such as in gas–solid systems. Ultimately, the motion of fluids involves either liquids, gases or mixtures. These involve various applications, ranging from medical sciences such as blood flow or transport of pollutants up to high-tech engineering and power systems. Extending the list of the individual applications does not make much sense, because the presence of fluids in technology and life sciences is ubiquitous. In any case, the subject of interest has been the fluid flow behaviour and its various interactions with its environment.

Fluid flow systems comprise different state variables, special components and obviously account for governing equations, describing the physics. For the solution of the complex flow phenomena, the literature usually refers to numerical solutions represented with the abbreviation CFD, i.e. computational fluid dynamics. In Chapter 1, Introduction to Multiphysics Modelling, it has been attempted to shed light on the systematic work flow for the successful multiphysics modelling and it has been elucidated that these fundamental steps are generally valid to any phenomenon, comprising multiphysics modelling.

A large range of engineering applications involve turbulent flow and simultaneous flow of multiple phases. Due to the vast field of applications, it is not possible to cover individual details of each topic. However, it will be useful to introduce the reader first to the commonly used fundamentals of single phase fluid flow systems. This helps the multiphysics analyst to become familiar with the fundamentals and handling of fluidic domains. The reader will become an insight to those advanced interacting fluid flow systems in upcoming chapters. The current chapter targets to draw attention on the correct use and solution of fluidic zones in multiphysics problems.

One of the most abundant fluid flow problems observed in practice is the application of single phase fluids [10–12]. Therefore quite a large range of multiphysics problems fall into this class [13–17]. This kind of flow can also include multiple fluid constituents such that it can either form a *reacting flow* [18–25] or be a *nonreacting flow*. In any case, one single fluid medium phase is formed at a scale well above the molecular level that is being transported through a domain, i.e. *internal flow* or such that an object is surrounded by the medium. In this case, the term *external flow* can be used.

In general, there are various variables that are of interest in a flow field problem. Pressure drop [26–29], velocity or integral properties such as lift and drag [30–33] are in various engineering fields of concern. Despite it is not possible to cover all of the fluid dynamics topics within the scope of this book, the chapter will point out important fluid flow facts that will be useful whilst modelling multiphysics.

As it is well known from basic fluid dynamics lectures, the solution for fluid flow systems may handle *laminar flow* [34,35] with ease and for simple geometries such as pipes and ducts in particular [36–40], but complex flows showing *turbulence* [41–45] require the use of more detailed and complex *turbulent flow* models

[46,47]. These are required to be solved in addition to the typical governing equations such as conservation of mass and momentum. For various applications of turbulent flow, there have been many different models developed so far [48–54].

In particular, the use of these models are indispensable in situations, where complex vortex dynamics [55], or surface wall treatments [56], reacting flow are of concern. Most of the turbulence models are derivatives of the so-called RANS—*Reynolds Averaged Navier–Stokes* models and their subcategories [57–63] that focus on the computation of the Reynolds stresses.

Another category the so-called LES—*Large Eddy Simulation* [64–67] and the hybrid variants of model categories [68–70] are also in modern turbulent flow modelling very common. Despite that no turbulent flow model is universal applicable, the standard turbulence models yield reasonable mean flow characteristics that are pretty helpful for many practical engineering problems. The expensive advanced models are superior in regard of offering predictive fidelity, which some specific applications require. The reader should refer to those multiequation models in special cases such as in investigating mean rotational effects or in smoke development in a fire for example, where a fine tuned model may be required. Hence, commonly used two equation models such as the k-omega or k-epsilon and their variants give for an initial description of turbulence, satisfactory results that can be fine tuned using more advanced models.

## 2.1 Governing Equations

The governing equations that describe the fluid flow are based on two main physical principles. These involve the conservation of mass and the application of Newton's second law of motion that expresses ultimately, the balance of momentum. The validity of the Newton's second law of motion requires that the fluid motion is within the confines of the continuum hypothesis that allows to define fluid particles; thus the fluid motion is considered on a macroscopic level. This enables the description of velocity and pressure.

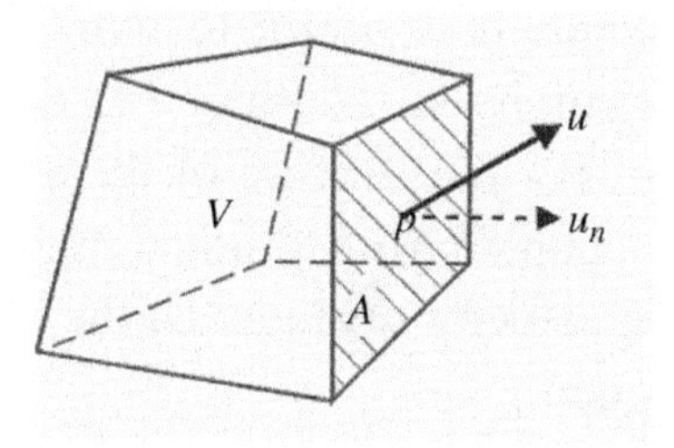


**Figure 2.1** Representative control volume ($Cv$).

### 2.1.1 Conservation of Mass Principle

The first important principle in a fluid flow is the conservation of the mass, which is usually named as the continuity equation. The conservation of mass principle implies that the rate of accumulation of mass inside a control volume and the net rate of outflow of the amount of mass across the control surface equal to zero.

To understand this, it is helpful to start with the control volume approach, enabling a simple integral analysis that will help the reader to understand the physics easier. The simple control volume or 'cell' of a fluid with volume $V$ is assumed (Fig. 2.1). The fixed mass of the fluid inside the cell would be the product of the volume $V$ with its density $\rho$ (Rho). The density or volume may change with time. But in any case, the mass inside must be unchanged, as to conserve the mass.

The reader can imagine this analogue to a balloon filled with hot air. When the balloon starts to exchange heat with its cooler surrounding air, what will happen is that the air inside the balloon will lose energy and the filled air becomes colder. The decrease in temperature will result in an increase in fluid density. The volume will change and the balloon will start to shrink in volume. But the mass of air inside the tight balloon will retain the same. Mathematically, this can be expressed as:

$$\frac{dm}{dt} = \frac{d}{dt} \int_{Cv(t)} \rho dV = 0 \tag{2.1}$$

When the time rate of change of mass of the control volume is related to the density and its volume, it becomes clear that (keeping in mind the balloon) the mass change must be zero.

In fluid flow problems, it is useful to work with velocity fields that gives us the opportunity to consider the control volume as an arbitrary fluid element. For one surface of the same considered control volume entirely occupied with fluid, designated with $S$ and an area of $A$ and the fluid velocity given as $U$, the mass flux across this surface can be determined.

The mass flux ($\varphi_f$), which is the rate of mass flow per unit area can be expressed in its integral form as:

$$\int_{S(t)} \rho U.ndA = 0 \tag{2.2}$$

$\varphi_m = \rho U_n A$, assuming a vector description for easy understanding and the net rate at which mass is flowing outwards ($n$ indicating the normal component, $\rho$ the density of the fluid, $S$ the surface and $A$ the area of the surface) through all the surfaces and the accumulated mass inside the volume must yield to zero, thus:

$$\int_{Cv(t)} \frac{\partial \rho}{\partial t} dV + \int_{S(t)} \rho U.ndA = 0 \tag{2.3}$$

Using the Gauss's theorem, the surface integral in Eq. (2.2) can be expressed in volume form using the Nabla operator as:

$$\int_{S(t)} \rho U.ndA = \int_{Cv(t)} \nabla.\rho U dV \tag{2.4}$$

simplifying in Eq. (2.3) will yield to:

$$\int_{Cv(t)} \frac{\partial \rho}{\partial t} + \nabla.\rho U dV = 0 \tag{2.5}$$

But when we consider that the control volume chosen has been an arbitrary one, but laying entirely in fluid as we are allowed to do so within the continuum hypothesis, then the integrand must be zero anywhere in the fluid. Because the negative and positive contributions will cancel each other, thus approving the assumption. Hence, the integral sign may vanish and the differential form can also be obtained at all points in the fluid.

$$\frac{\partial \rho}{\partial t} + \nabla.\rho U = 0 \tag{2.6}$$

The differential form is one of the fundamental equations in fluid dynamics and variants will not be further handled in this context.

### 2.1.2 Conservation of Momentum

The conservation of momentum or sometimes expressed as equation of motion of a fluid states in general that the time rate of change of momentum of a selected section of a fluid is equal to the sum of all the forces acting on the fluid section or the defined system. The reason why this formulation has been chosen more appropriate instead of directly employing the Newton's second law of motion in its usual form, i.e. $F = ma$ is, the fluid cannot be considered as if existing point masses that are independent of time. However, considering that the mass is a product of volume and density, it is possible to stay with the previous dependent variables. This enables to consider that the force per volume equals to the change of momentum per volume.

When the control volume is considered *totally* as fluid and our considered control volume viewed as *Eulerien*, the derivative of the velocity can be used as to present the acceleration term and to calculate the time-rate of change of momentum.

$$\frac{D}{Dt}\int_{Cv(t)} \rho U dV \equiv \{\text{time rate of change of the momentum vector}\} \tag{2.7}$$

Considering the body forces that act on the entire control volume, as well as the surface forces acting on particular surfaces within the $Cv$, an expression for the fluid motion can be determined such as:

$$\frac{D}{Dt}\int_{Cv(t)} \rho U dV = \int_{Cv(t)} F_B dV + \int_{S(t)} F_S dA \tag{2.8}$$

but the description of the forces are required to utilise the formulation in calculations. By moving the differentiation inside the integrals using the Reynolds transport theorem and following with a Gauss's theorem applied to the surface integral section, it is possible to express the right side as:

$$\int_{Cv(t)} \frac{\partial \rho u}{\partial t} + \nabla.(\rho u U) dV \tag{2.9}$$

It is possible to differentiate with the product rule the term:

$$\nabla.(\rho u U) = U.\nabla(\rho u) + \rho u \nabla.U \tag{2.10}$$

Note that by considering incompressible flow and constant density, further simplification can be performed as to obtain:

$$\nabla.(\rho u U) = \rho U.\nabla U \tag{2.11}$$

hence, the governing equation of momentum can be given as:

$$\frac{D}{Dt}\int_{Cv(t)} \rho u dV = \int_{Cv(t)} \rho\frac{\partial u}{\partial t} + \rho U.\nabla u dV \tag{2.12}$$

One important technical application appearing a lot in fluid dynamics problems is the use of perforated, voided materials or components. In such cases, the flow path along which the fluid is transported inside a domain can comprise void fractions such as in a porous object. These kind of materials are usually treated according to the continuum theory of porous media. Geophysics, textiles, energy and civil engineering are only a few of various areas, utilising these kind of materials.

Many analysts try to make use of the porous media assumption, as to simplify their problem. The modelling of porous materials and components has been a topic since many years. Over the years, personal experience shows that many of the students, engineers and scientists are eagerly willing to utilise the continuum approach of porous media. However, it is not rare that the assumptions they make is leading to ambiguous results.

They attribute these results or the deviations from experimental measurements to either the errors in experiments or to numerical uncertainties. However, it is very common that these uncertainties are due to the inaccurate assumptions or chosen mathematical models. Therefore attention has been drawn to some important aspects on the CFD modelling of these materials and simplifying components or materials geometrically to a continuum, as to make an appropriate use of the porous media theory.

First of all, the reader is briefly introduced to what a porous media is and physically how it has been utilised in fluid flow modelling. Thereafter, it is important to determine what kind of mathematical model depicts the real physical behaviour and most importantly to understand whether or not the assumption is valid for the solution of the specified problem.

Many engineering applications such as catalyst beds, filters and fuel cells utilise porous materials, thus involve the flow through porous media. For the purpose of this book, a solid containing holes or voids, either connected or nonconnected, dispersed within it in either regular or random manner will be classed as a porous material. Provided that such holes occur relatively frequently within the solid [71]. It should be noted that fluid flow *through* a porous material can only be considered if at least some of the pores are interconnected.

A typical practical situation is where the analyst assumes that the flow through the porous medium is obeying the *Darcy* law [72–85]. However, the interpretation and validity of the approach needs to be justified. There have been many practical cases where the

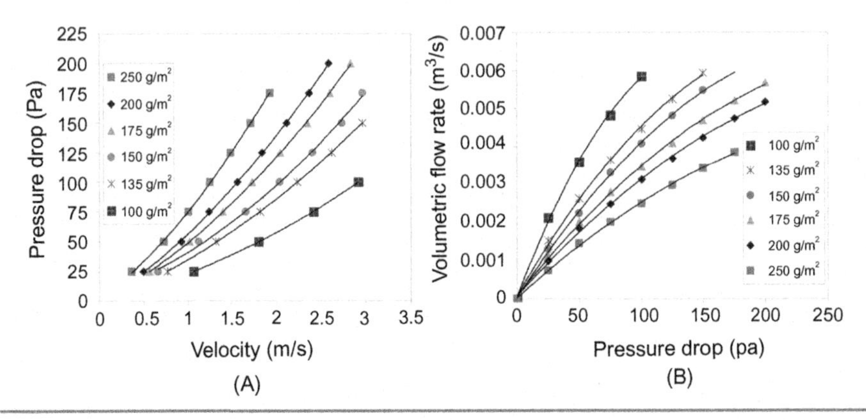


**Figure 2.2** (A) Pressure drop versus air velocity as a function of fabric weight density, (B) flow rate versus pressure drop as a function of fabric weight density [86].

experimental measurements regarding the flow through a certain porous material show that a nonlinear relationship between the pressure drop and the volume flow rate is observed. The Darcy theory, however, is only valid for the linear section of this relation.

Thus, the nonlinear part is not considered. Therefore from scratch the user implements errors into the analysis at the initial stage of the set up. These errors will be reflected in the results, as well. The following sample is exemplifying this case. Fig. 2.2 illustrates the pressure drop versus volume flow rate of the air flow through a technical geotextiles sample [86]. Analysing the relationship between the volumetric flow rate and the pressure drop for different fabric area densities illustrates this nonlinearity clearly. The decrease of air permeability is visible as the fabric weight increases (Fig. 2.2B). The reason for this is the porosity and the fabric weight density gradient within the web. This is due to the different interconnected pores through the fabric thickness of the fibrous web.

When this behaviour would be mathematically described as linear and implemented in the numerical calculation, the solver will assume this linearity as dictated by the user. However, the results will be manipulated. In this example the experimentally validated simulation model utilised the mathematical description that accounted for the Forchheimer term [87], describing the nonlinear section, as well [88–95].

Mathematically this is performed such that for the solution of the momentum equation, a momentum source term is added to the Navier–Stokes equations. The source term is comprising two parts, i.e. a viscous loss term and an inertial loss term, expressed as:

$$S_i = -\left(\frac{\mu}{k} + \frac{\beta \rho_a}{2}\left|\vec{u}\right|\right)\vec{u} \tag{2.13}$$

This momentum sink (mind the minus sign) $S_i$ contributes to the pressure gradient in the porous cell, creating a pressure drop for a given thickness of the porous layer. The first term describes the viscous loss term where $k$ (not to confuse with thermal conductivity) is the air permeability, $\mu$ the dynamic viscosity of the fluid, and let $\vec{u}$ be the velocity vector. $1/k$ in this equation is denoted as the constant viscous resistance factor.

The second term is expressing the Forchheimer regime that describes the transition from laminar flow to low turbulence. It refers to the inertial effects. $\beta$ and $\rho_a$ present the inertial resistance coefficient and the air density, respectively. $\phi$ stands for the porosity of the material. Hence, the momentum equation for this situation ends up such as expressed in Ref. [87]:

$$\begin{aligned}\rho_a \frac{\partial(\phi\, u^{\rightarrow})}{\partial t} + \rho_a(\phi\, \vec{u}\, .\nabla)\, \vec{u} &= -\,\phi \nabla p + \mu \nabla^2(\phi\, \vec{u}) \\ &\quad - \left(\frac{\mu}{k} + \frac{\beta \rho_a}{2}\left|\vec{u}\right|\right)\vec{u}\end{aligned} \tag{2.14}$$

The most reliable way to define the resistance coefficients of a porous structure is to employ tabulated experimental data. Experimental data for the porous components in the form of pressure drop against velocity through the component enables to extrapolate the data, in order to determine the coefficients for the used materials. For a simple homogenous porous medium, the pressure drop is proportional to the fluid velocity (or velocity squared) in the cell and can be formulated as

$$\Delta p = \mathrm{au} + bu^2 \tag{2.15}$$

Hence, the coefficient, $a$, is describing the Darcy regime and the coefficient $b$ refers to the Forchheimer regime. A curve fit of the data is plotted to create a trend line or curve fit through these points, yielding the Eq. (2.15), i.e. a polynomial in velocity, so taking the curve coefficients:

$$b = i_r \frac{1}{2} \rho_a \Delta n \tag{2.16}$$

with $i_r$ the initial resistance factor and $\rho_a$, $\Delta n$ the air density for the specified temperature and the porous layer thickness, respectively.

Likewise, with

$$a = \frac{\mu}{k} \Delta n \tag{2.17}$$

where $\mu$ is the viscosity of the air, and $1/k$ the viscous resistant factor, the coefficients are derived.

Another important feature to draw attention when using this approach is to make accurate assumptions during the computational modelling stage. To computationally model a continuum structure has been an approach that many scientists and engineers employ. However, defining a porous physical geometry as a continuum structure requires great care. Thus the feature 'understanding the flow process' is emphasised.

The mathematical interpretation and accurate description of the flow behaviour has been one issue so far. The following practical case is an invaluable example that aims to alert the inexperienced people to the problems associated with insufficient understanding of the physical parameters and its effect on the numerical results. Without a deep understanding of the real flow process, the analyst may stick on inadequate assumptions and produce confusing results. Ultimately, these kind of errors will tarnish the reputation of the simulation technology and its powerful support in research and product development. In reality, it has been the assumption, thus the user being responsible for the inconvenience caused.

In the following example, the porous media approach has been used with the aid to determine the flow distribution inside a fuel cell component. A fuel cell interconnector plate is one major component containing long channels, where the understanding of the mal flow distribution is of paramount importance. There have been many modelling cases where this kind of complex geometries has been assumed as porous media. It is important to know which approach to choose and to be aware of its effects.

When considering the numerical modelling of a porous material or component, defining and implementing the flow resistance is required. The flow domain is geometrically created as a continuum that is specified as a porous, i.e. pseudofluid domain. In the present example, the analyst would usually assume that the flow through the narrow channels is the major part in the flow of an interconnector plate. Thus the flow is dominated in two directions where the flow resistance in the third dimension can be ignored and modelled such that a large resistance in this direction is implemented.

This will direct the flow of course in a certain flow direction. The flow resistances for the directions are usually determined experimentally. Flow measurements are used to determine the pressure drop and deriving parameters for the flow resistance. However, the flow field in 3D is not utilised for this process. On the other hand if the channels would have been modelled fully resolved, these parameters would not be required. Whilst, investigating and justifying the validity of the approach resulted in a very interesting comparison.

The first contour plot in Fig. 2.3 illustrates the pressure distribution of the fluid domain inside the component, by modelling the channels fully resolved. The porous media model has been used as the second analysis. In practice, the results reveal a bad flow distribution. But there is more behind. The real profile of the flow shows a wavy distribution. The first attempts showed different results, initially. Why was there a discrepancy?

The answer lies in the assumptions chosen in the porous media approach. The infinite large flow resistance given in the particular

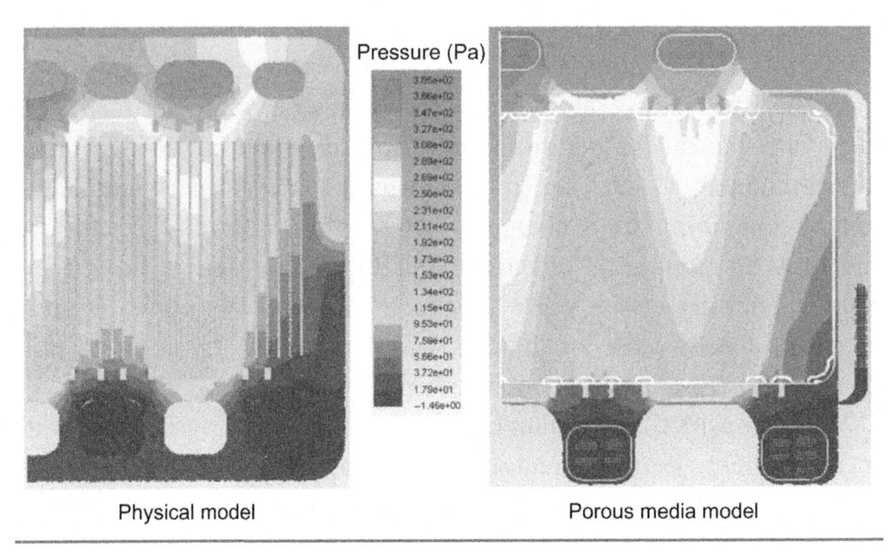


**Figure 2.3** Pressure distribution comparison using a physical model versus porous media model.

direction led to a directional flow that was responsible for the different distribution. The curiosity how to mimic the real profile resulted by trial and error to the same distribution after adjusting the particular flow resistance values such that both results could depict the similar distribution. By adjusting the flow resistance it was ultimately possible to show that under the specified resistances the flow approximated such a profile. Without knowing the real profile, an estimation would not be possible.

This suggests the careful use of porous media approach, as it may have drastic consequences in the design and predevelopment phase of a component. The multiphysics analyst is advised to choose physically resolved models in cases where the flow resistances cannot be described appropriately.

## 2.2 Considering Turbulence in Multiphysics Modelling

In the previous section, the basic governing principles regarding mass and momentum were developed and applied in conjunction with severe assumptions to some flow situations. However, the flow of a fluid loses its stability above a certain Reynolds number ($UL/\nu$ where $U$ and $L$ are characteristic velocity and length scales of the mean flow and $\nu$ is the kinematic viscosity).

This stability is in practice experimentally demonstrated where neutrally buoyant dye is injected into the flow path. For flow rates under a 'certain' amount, the *streakline* of the dye will retain a well-defined line as it flows along the surrounding fluid. As the flow rate 'increases', the dye streakline starts to fluctuate in time and space and this progression leads to intermittent bursts of irregular behaviour that starts to appear along the streakline. Ultimately, when the flow rate 'increases further', the dye streak becomes blurred and spreads across the entire volume in a random manner (Fig. 2.4).

It is this distinguishing characteristics of the flow that has been denoted as laminar, transitional and turbulent. The relative quantity of the flow rate is with this categorisation also been used as relative to a reference quantity, i.e. the Reynolds number that describes the analyst the ratio of the inertia to viscous effects in the flow. The transition between these three characteristics is strongly affected by pressure gradients, disturbance levels and the surface roughness, as well as the presence of heat transfer.

As the flow gets turbulent, conserved quantities of the flow get stirred. Thereby, the contact between fluid particles with different momentum is increased and with diffusion, mixing occurs. The effects produced by turbulence may in some applications be useful;

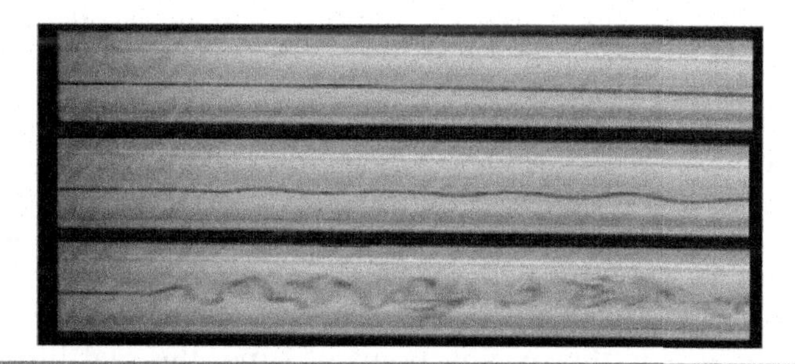

**Figure 2.4** Time dependence of fluid velocity measured at a point: Laminar (top), transitional (middle) and turbulent (bottom).

particularly, intense mixing is in chemical species transport and heat transfer desired.

However, it should be noted that when mixing increases, frictional forces also increase, thus more power is required in some applications to produce work via fluid motion. The point of instability and onset to the transition from laminar up to turbulent depends on the amplification of the disturbances. This may involve the formation of local turbulent spots such in flow over plates where boundary layers occur or flows where points of inflexions arise such as in complex jet flows.

But in common, the transition starts from small disturbances that leads to rotational shapes and small scale motions that ultimately merge together to build up into a full turbulent flow. Thus the transition process is very much dependent of the case. The belief that at Reynolds numbers above 2300 always turbulent flow occurs, is not valid. There has been great progress in understanding this kind of transitions by investigating thoroughly various geometries in flow streams, both experimentally and numerically.

The crucial differences in visualisation of laminar and turbulent flows are the appearance of *eddies* of various length scales in turbulent flows. The larger coherent ones are different in each flow, whereas the smaller eddies are similar in most of the flow. In the large eddies, most of the energy is stored and is transferred to the smaller eddies. In the smallest eddies, the turbulent energy is converted through viscous dissipation into internal energy. To track these eddies numerically requires also the direct solution of the time-dependent Navier–Stokes equations, which also demands high computer performance.

Thus time-averaged fluid properties are used to get satisfactory information about the turbulence behaviour. This approach is usually named as the RANS approach. It should be noted that in this approach, many details regarding instantaneous fluctuations are lost. Fig. 2.5 illustrates this difference.

The difference is that in any point of time, the real-life instantaneous velocity comprises the time-averaged velocity, in addition to

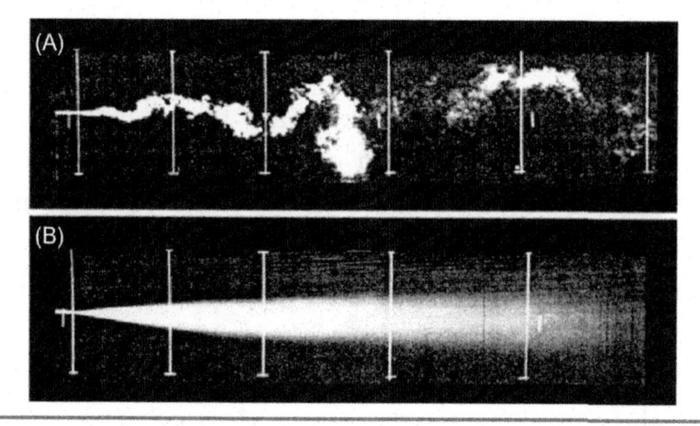


**Figure 2.5** Flow visualisation of instantaneous (A) and time-averaged (B) turbulent plume downstream. *Photograph courtesy of the U.S. EPA/ NOAA Fluid Modelling Facility.*

the fluctuating velocity. It is the vector components of this fluctuating velocity, which make up the turbulent kinetic energy $k$, expressed as:

$$k = \frac{1}{2}(\overline{u^2} + \overline{v^2} + \overline{w^2}) \tag{2.18}$$

Modern anemometer devices such as laser Doppler and variants are widely used for mean flow and turbulence measurements.

As many fluid flow systems show significant turbulent characteristics, the turbulent flow regime and consideration in multiphysics modelling is not just of theoretical interest. This complexity can best be handled with numerically solved multiphysics modelling. Hence, the section gives a brief introduction to the modelling of turbulent flow.

When accounting for turbulence in multiphysics modelling, the classical fluid dynamics models use the time-averaged approach, where only the mean turbulence effects on the mean flow are sought after. These are simple and economical to utilise on most of the computers. At the beginning of the chapter, the advanced time-dependent approaches such as the LES and Direct numerical simulation (DNS) have been mentioned. The DNS has been widely used in academia to extract verification data for the time-averaged approaches, as they are very large computer models, thus very expensive to run. The numerical mesh and time steps are very demanding.

As the main effects of turbulence are dominated by the larger eddies due to their higher amount of energy, the utilisation of LES simulations has been an alternative to the DNS. As both approaches are too costly in considering general purpose multiphysics analyses, the interested reader may refer for more details to the mentioned literature.

Instead, it is intended to draw more attention to the classical models presently used by far the most. Most of them are validated and appear in most complex multiphysics applications to be satisfactory to account for the turbulences and predict certain quantitative properties. It should be noted that in multiphysics analyses there will be more effects that interact together. As the multiphysics applications require a lot of computational resources, it is wise to benefit from the most effective solution. To improve the understanding of the classical approaches, the RANS models have been chosen, as they have a wide range of application.

They assume that an analogy between the viscous stresses and the Reynolds stresses exist on the mean flow. As the continuity equation receives no additional terms, the averaged form is analogue to the previously derived one. However, an additional task on the momentum equation is required.

The general form of the Reynolds-averaged momentum equations can be expressed as:

$$\rho\left(\frac{\partial \overline{u_i}}{\partial t}+\overline{u_k}\frac{\partial \overline{u_i}}{\partial x_k}\right)=-\frac{\partial \overline{p}}{\partial x_i}+\frac{\partial}{\partial x_j}\left(\mu\frac{\partial \overline{u_i}}{\partial x_j}\right)+\frac{\partial R_{ij}}{\partial x_j} \tag{2.19}$$

where $R_{ij}$ is the stress tensor, which appears due to the averaging procedure of the convective term on the momentum equation. This is the crucial term that needs to be modelled, so that the equation gets mathematically close. This is either handled with the so called eddy viscosity models or with the Reynolds stress models. The stress tensor is often modelled using common viscosity models that utilise the Boussinesq approximation. It can be expressed as:

$$R_{ij} = -\rho\overline{u_i'u_j'} = \mu_t\left(\frac{\partial\overline{u_i}}{\partial x_j} + \frac{\partial\overline{u_j}}{\partial x_i}\right) - \frac{2}{3}\mu_t\frac{\partial\overline{u_k}}{\partial x_k}\delta_{ij} - \frac{2}{3}\rho k\delta_{ij} \tag{2.20}$$

It is the turbulent eddy viscosity $\mu_t$ that has been calculated by each turbulent model differently.

The most widely used turbulence models in engineering applications are the $k$-$\varepsilon$ turbulence model variants. They are robust and cost effective. The $k$-$\varepsilon$ models focus on the mechanism that effect the turbulent kinetic energy. Each field variable is defined as the sum of the fluctuation and the mean components. The standard $k$-$\varepsilon$ are one of the most widely used models [96]. They perform particularly well in confined flow that are parallel in channelised structures.

The models show difficulties in predicting flows with strong separation and swirls, rotating flows. The equation includes two models one for the turbulent kinetic energy and one for the viscous dissipation. These are used to express the velocity scale ($\vartheta$) and the length scale ($l$). They are representative for the large scale turbulence expressed as:

$$\vartheta = k^{0.5} \quad \text{and} \quad l = \frac{k^{1.5}}{\varepsilon} \tag{2.21}$$

All $k$-$\varepsilon$ models are based on the calculation of the eddy viscosity$\mu_t$, expressed as:

$$\mu_t = C\rho\vartheta l = \rho C_\mu\frac{k^2}{\varepsilon} \tag{2.22}$$

where the dimensionless constant $C_\mu$ is empirically determined as 0.09.

The standard $k$-$\varepsilon$ model uses the transport equation for the term $k$ in its most commonly used form as:

$$\frac{\partial(\rho k)}{\partial t} = -\nabla(\overline{U}\rho k) + \nabla\left(\frac{\mu_t}{\sigma_k}\nabla k\right) + P_t - \rho\varepsilon \tag{2.23}$$

and the transport equation for $\varepsilon$ can be expressed as:

**TABLE 2.1 Coefficients Used in Transport Equations**

| $\mu_t$ | $\sigma_k$ | $\sigma_\varepsilon$ | $C_{1\varepsilon}$ | $C_{2\varepsilon}$ |
|---|---|---|---|---|
| | 1.0 | 1.3 | 1.44 | 1.92 |

$$\frac{\partial(\rho\varepsilon)}{\partial t} = -\nabla(\overline{U}\rho\varepsilon) + \nabla\left(\frac{\mu_t}{\sigma_\varepsilon}\nabla\varepsilon\right) + C_{1\varepsilon}\frac{\varepsilon}{k}P_t - C_{2\varepsilon}\rho\frac{\varepsilon^2}{k} \tag{2.24}$$

the production term $P_t$, which is a function of the average velocity gradient is given as:

$$P_t = -\frac{\mu_t}{2}\left|\nabla\overline{U} + \nabla\overline{U^T}\right|^2 \tag{2.25}$$

The equations comprise five constants that are derived from various data fittings and are expressed in Table 2.1.

## 2.3 Practical Review Examples

As it has already been pointed out, multiphysics simulation requires experience in various disciplines and physical phenomena. However, to support and utilise this knowledge, the analyst is supposed to exercise as much as possible. This will provide great understanding of particular problems and analogue situations, as well as improve the confidence in applying and executing solutions for various multiphysics problems. As the book is intending to prepare the reader stepwise into the world of multiphysics, a brief practical support section is accompanying this chapter.

It is very important to start with these kind of simple problems as training and practice material. These usually have analytical solutions that give the reader the opportunity to understand the verification characteristics in multiphysics simulation. This grants a feeling about the solution in complex problems that will be demonstrated in ongoing chapters. Moreover, it is advisable that by changing values and parameters of the problem, the user can improve the instinct to interpret mesh quality and its effects on the solution.

Experience shows that many students and new users to multiphysics simulation use various simulation codes. This may be due to availability, experience or any other source of reason. Everyday, new providers or versions of existing software come into the market. In order to provide software independent information, it is aimed to describe or solve few simple problems in the fundamental first four chapters in this book such that any reader can easily draw and apply the described problems.

This will not only enable the reader to use their own favourable software, but also mitigate to advertise a certain code or provider, as this is not the purpose of this book. The user should be free of software constrains and focus on the problem basics. This is required to build a strong foundation and to proceed further.

In the following example, a multiphysics model is used to simulate the air flow through a channel with a rectangular settling and a solid obstacle such as a tip section of a cylinder. Air flows from the left inlet area into the channel and leaves the domain from the right, flowing over the obstacle. The turbulent flow solution is calculated using the standard $k$-$\varepsilon$ model. The model assumes symmetry. The inlet velocity of 15 m/s and the outlet atmospheric pressure is known.

A structured grid is used for the problem and all boundary conditions are labelled on the computational domain. The structured grid has been chosen to be a multiblock domain. For this purpose, the analyst has to create blocks inside the half rectangular settling. The block structure mitigates skewed cells and the cells are clustered at the vicinity of the obstacle. These will be specified as interior zones (Fig. 2.6).

The simulation results show in detail the locally separated flow region over the tip region of the obstacle. These fluctuations result in association with the unsteady wall pressure the obstacle is subjected to. The region of separated flow over the tip defines the largest scale eddies, which are present in the flow. It is of very common practice to validate such profiles or pathlines using smoke flow visualisations. Such an example has been illustrated in Fig. 2.7.

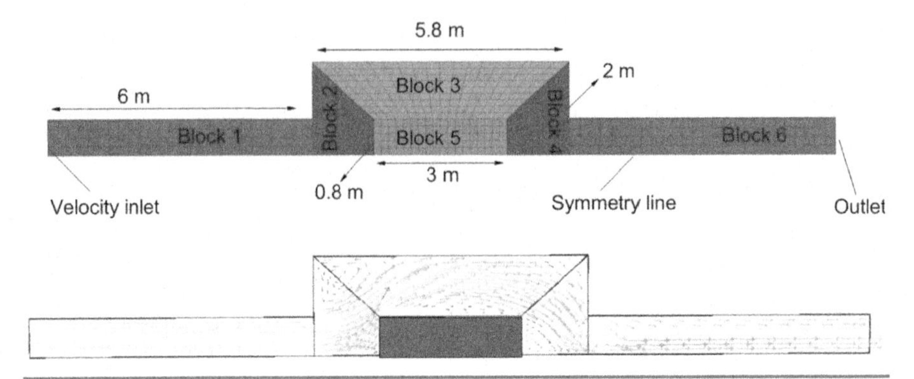


**Figure 2.6** Sample channel geometry and simulated velocity vector field.

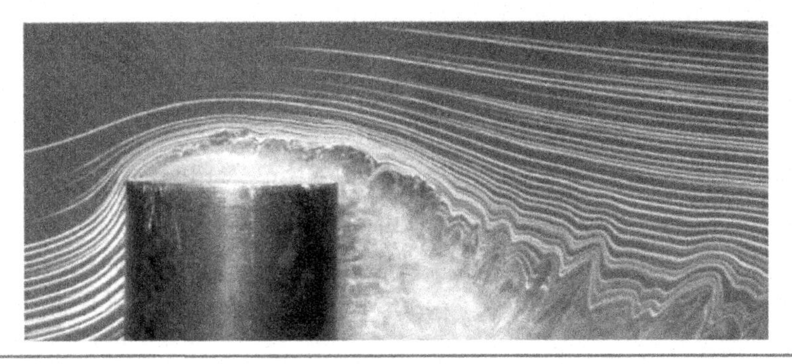

**Figure 2.7** Smoke wire flow visualisation of the flow over the end-cap of a right cylinder [97].

The next example demonstrates a multiphysics analysis, comprising four different material layers that are stacked on each other. The cell zones of each layer are presented as porous fluid zones, allowing air as the fluid medium to flow through each model component. Porosities have been chosen such that the layers decrease in porosity in the direction from the flow inlet to the flow outlet region. It should be noted that for more accurate simulations of porous media flows, the analyst should use the true or physical velocity throughout the flow field, rather than the superficial velocity.

Therefore the value that enters the system before multiplied by the porosity is used. An inlet flow velocity of 3 m/s has been applied. The outflow boundary is modelled by setting the pressure to zero. It is assumed that the porous region always models a control space and that only the air contained in the process can leave the control space. Furthermore, it is assumed that the pores are statistically distributed. The flow is considered as incompressible, laminar, therefore the no-slip conditions are applied ($u = v = 0$).

The face zones inside the computational domain represent the interior zone. The model is assumed to be a section of a radial system, thus rotational periodic boundary conditions are specified. A regular structural numerical grid utilising 2D quadrilateral cells has been set up.

The staggered mesh arrangement leads to a strong coupling between the velocities and the pressure, enabling faster solution times. The reason is the implicit relationship that exists between a cell and its neighbours in a regular numerical grid, which permits data to be found easily. The model details and the solved pressure and velocity distributions are illustrated in Fig. 2.8.

The demonstrated air velocity distribution of the layers shows that the velocity through Layer_D appears to be the highest, whereas the upper layers reveal lower velocity values. This is due to the porosity differences among the components. Layer_D has been assumed to have the lowest porosity, thus forces the air to flow faster between its pores. In comparison, Layer_A with a higher porosity allows the air to pass through with less resistance, explaining the lower velocity values. The pressure distribution has been illustrated as a reminder to the analyst. As it can be noticed, in an incompressible flow like in this case, the pressure values decrease at the regions where the velocity increases and vice versa.

The last example in this chapter has been chosen based on a typical fundamental fluid mechanics problem that has been of practical importance. The flow over a cylinder problem demonstrates an upstream length of 0.75 m and a downstream length of 40 times the radius of the cylinder, i.e. 2 m. The width of the flow domain is 50 times the radius of the cylinder. The problem has been depicted in Fig. 2.9.

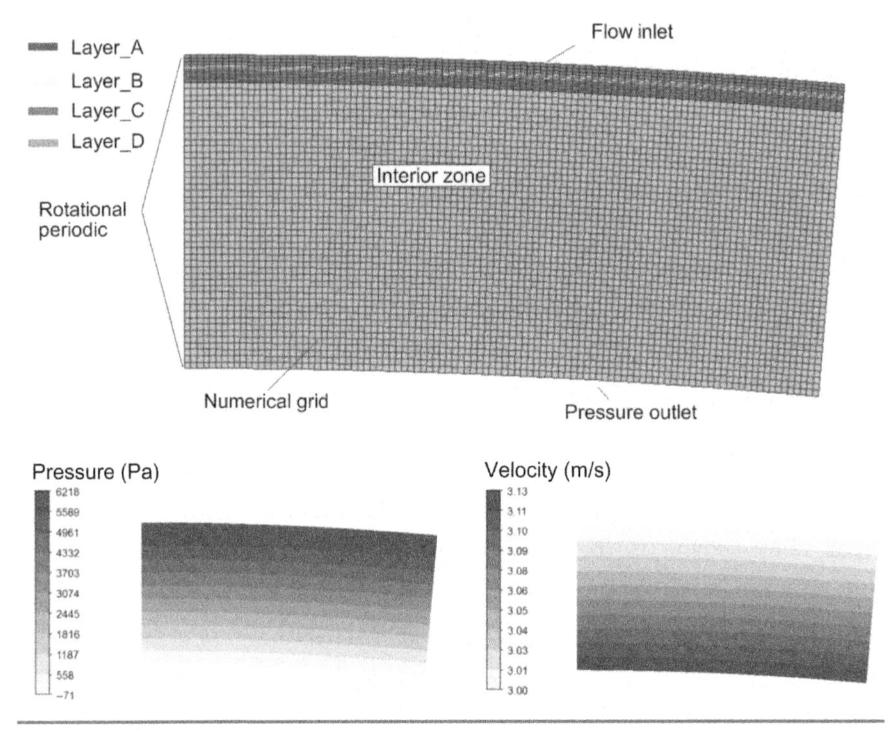


**Figure 2.8** 2D multiphysics model details: model description, as well as distributions of the pressure and velocity fields.

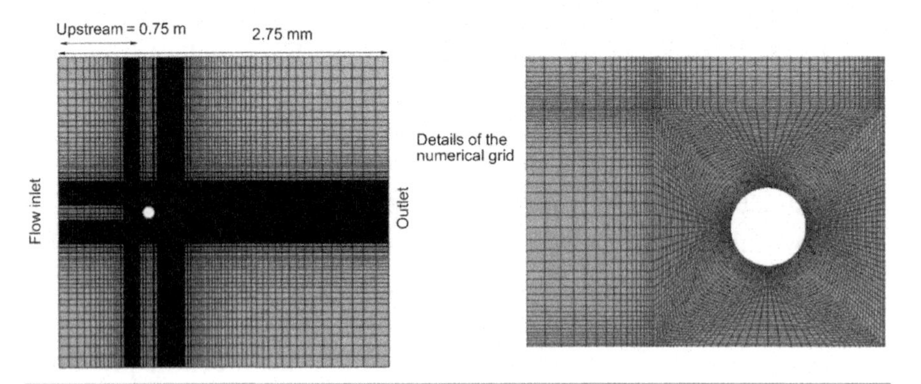


**Figure 2.9** Details of the used numerical model.

The analyst should note that again multiblock regions are created as to efficiently generate a numerical grid; particularly, to capture the effects at the upstream and downstream regions. The experienced analyst will know that a first cell height placed around a particular

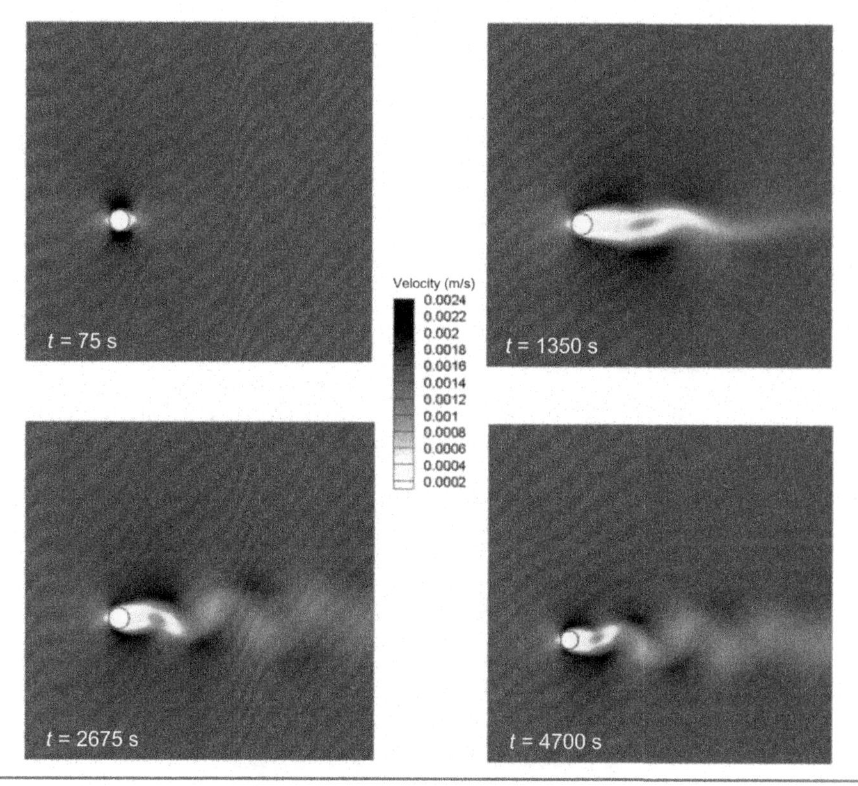


**Figure 2.10** Transient velocity distribution of the flow over the cylinder.

region is utilised in turbulent modelling. However, a used Reynolds number of around 150 enables here the use of a laminar model, thus the constraints on height of the first cell is not a prerequisite.

Important is that the gradients are adequately resolved. Therefore attention has been given to apply a boundary layer around the cylinder of multiple layers. The experienced analyst will use a more accurate approach and consider the theoretical calculations for the cell intervals used for meshing the domain. However, most of the preprocessors utilise sufficient algorithms to provide the new analyst with an adequate numerical grid. For the solution of the problem, a transient analysis has been used.

Fig. 2.10 shows the velocity distribution predicted for certain time instants. It can be noticed that an unstable flow is present and

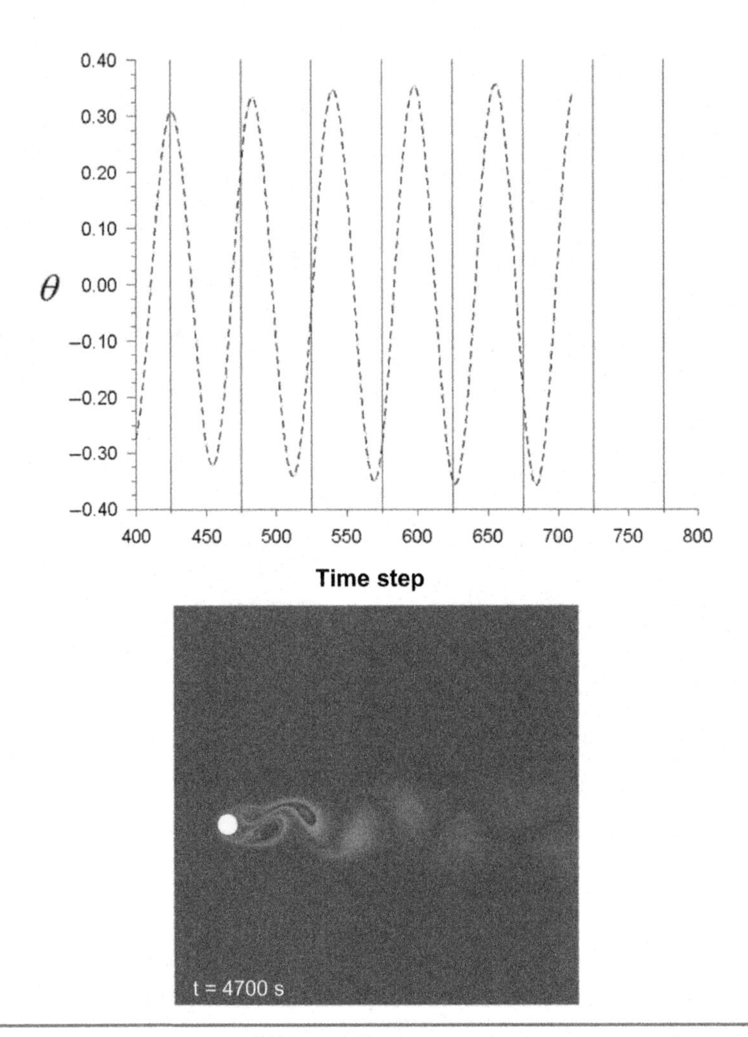


**Figure 2.11** Variation of lift coefficient with time step and an example of the vorticity field for $t = 4700$ s.

flow separation occurs. This results in an asymmetric behaviour at the downstream side that is expected, according the theory of stability. The wake behind the cylinder is visible where the flow is slow. The highest speed is recognised above and below the cylinder, where less resistance is present.

Consequently, another interesting output for the analyst will be the lift coefficient history, acting on the body. This distribution is used to become an understanding of the vortex shedding frequency behind the cylinder. This will help to determine the instabilities that are important in fluid flow systems. For this purpose, the analyst needs to plot the time period of the flow oscillation, using the lift coefficient history over the cylinder (Fig. 2.11). Considering the approximate step time values of 540 (3600 s) and 600 (4000 s), the diameter of 0.1 m and the velocity of 0.0015 m/s, a Strouhal number of 0.167 is predicted. This value is in good agreement with the literature, revealing values of 0.170–0.175 [98].

## References


[1] Osswald T, Rudolph N, Osswald T, Rudolph N. 3 – Generalized Newtonian fluid (GNF) models. In: Underst. Plast. Rheol. 2015. 59–99. Available from: http://dx.doi.org/10.3139/9781569905234.003.

[2] Labrosse G, Kasperski G. Newtonian fluids and thermohydraulics. Encycl Math Phys 2006;492–502. Available from: https://doi.org/10.1016/B0-12-512666-2/00169-3.

[3] Holland FA, Bragg R, Holland FA, Bragg R. 2 – Flow of incompressible Newtonian fluids in pipes and channels. In: Fluid Flow Chem. Eng. 1995. 70–95. Available from: http://dx.doi.org/10.1016/B978-034061058-9.50004-9.

[4] Michael Lai W, Rubin D, Krempl E. Newtonian viscous fluid 6. Introd to Contin Mech 2010;348–426. Available from: https://doi.org/10.1016/B978-0-7506-8560-3.00006-2.

[5] Oldroyd JG. Chapter 16 – Non-Newtonian flow of liquids and solids. Rheology 1956;653–82. Available from: https://doi.org/10.1016/B978-0-12-395694-1.50022-1.

[6] Zienkiewicz OC, Taylor RL, Nithiarasu P, Zienkiewicz OC, Taylor RL, Nithiarasu P. Chapter 5 – Incompressible non-Newtonian flows. In: Finite Elem. Method Fluid Dyn. 2014. 163–94. Available from: http://dx.doi.org/10.1016/B978-1-85617-635-4.00005-4.

[7] Chhabra RP, Richardson JF, Chhabra RP, Richardson JF. Chapter 1 – non-Newtonian fluid behaviour. In: Non-Newtonian Flow Appl. Rheol. 2008. 1–55. Available from: http://dx.doi.org/10.1016/B978-0-7506-8532-0.00001-9.

[8] Lai WM, Rubin D, Krempl E, Lai WM, Rubin D, Krempl E. Chapter 8 – Non-Newtonian fluids. In: Introd. to Contin. Mech. 2010. 443–509. Available from: http://dx.doi.org/10.1016/B978-0-7506-8560-3.00008-6.

[9] Vossoughi S. Flow of non-Newtonian fluids in porous media. Rheol Ser 1999;8:1183–235. Available from: https://doi.org/10.1016/S0169-3107(99)80017-3.

[10] Huinink H. Single phase flow. Fluids in porous media. Morgan & Claypool Publishers; 2016. p. 6–9. Available from: https://doi.org/10.1088/978-1-6817-4297-7ch6.

[11] Abou-Kassem JH, Ali SMF, Islam MR, Abou-Kassem JH, Ali SMF, Islam MR. Chapter 7 – Single-phase flow equation for various fluids. In: Pet. Reserv. Simulations; 2006. 207–282. Available from: http://dx.doi.org/10.1016/B978-0-9765113-6-6.50013-7.

[12] Hernández A, Hernández A. Chapter 2 – Single-phase flow. In: Fundam. Gas Lift Eng. 2016. 25–79. Available from: http://dx.doi.org/10.1016/B978-0-12-804133-8.00002-6.

[13] Kandlikar SG, Garimella S, Li D, Colin S, King MR, Colin S. Chapter 2 – Single-phase gas flow in microchannels. In: Heat Transf. Fluid Flow Minichannels Microchannels; 2014. 11–102. Available from: http://dx.doi.org/10.1016/B978-0-08-098346-2.00002-8.

[14] Wu Y-S, Wu Y-S. Chapter 10 – Multiphase fluid and heat flow in porous media. In: Multiph. Fluid Flow Porous Fract. Reserv. 2016. 251–264. Available from: http://dx.doi.org/10.1016/B978-0-12-803848-2.00010-6.

[15] Kandlikar SG, Garimella S, Li D, Colin S, King MR, Kandlikar SG. Chapter 3 – Single-phase liquid flow in minichannels and microchannels. In: Heat Transf. Fluid Flow Minichannels Microchannels; 2014. 103–74. Available from: http://dx.doi.org/10.1016/B978-0-08-098346-2.00003-X.

[16] Chin WC, Chin WC. Chapter 4 – Steady, two-dimensional, non-Newtonian, single-phase, eccentric annular flow. In: Manag. Press. Drill. 2012. 127–81. Available from: http://dx.doi.org/10.1016/B978-0-12-385124-6.00004-6.

[17] Abou-Kassem JH, Ali SMF, Islam MR, Abou-Kassem JH, Ali SMF, Islam MR. Chapter 2 – Single-phase fluid flow equations in multidimensional domain. In: Pet. Reserv. Simulations; 2006. 7–41. Available from: http://dx.doi.org/10.1016/B978-0-9765113-6-6.50008-3.

[18] Murthy SNB. Turbulent mixing in non-reactive and reactive flows: a review. In: Murthy SNB, editor. Turbul. Mix. Nonreact. React. Flows. Boston, MA: Springer New York; 1975. p. 1–84. Available from: https://doi.org/10.1007/978-1-4615-8738-5_1.

[19] Anantpinijwatna A, Sin G, O'Connell JP, Gani R. A framework for the modelling of biphasic reacting systems. Comput Aided Chem Eng 2014;34:249–54. Available from: https://doi.org/10.1016/B978-0-444-63433-7.50026-2.

[20] Grossman B, Cinnella P. The computation of non-equilibrium, chemically-reacting flows. Comput Struct Mech Fluid Dyn 1988;79–93. Available from: https://doi.org/10.1016/B978-0-08-037197-9.50013-9.

[21] Marchisio DL, Fox RO. Reacting flows and the interaction between turbulence and chemistry. Ref Modul Chem Mol Sci Chem Eng 2016. Available from: https://doi.org/10.1016/B978-0-12-409547-2.11526-4.

[22] Oran ES, Boris JP. New directions in computing reacting flows. Comput Struct Mech Fluid Dyn 1988;69–77. Available from: https://doi.org/10.1016/B978-0-08-037197-9.50012-7.

[23] Boyadjiev CB, Babak VN, Boyadjiev CB, Babak VN. PART 3 – Chemically reacting gas-liquid systems. Non-Linear Mass Transf Hydrodyn Stab 2000;171–223. Available from: https://doi.org/10.1016/B978-044450428-9/50004-1.

[24] Carter JG, Cokljat D, Blake RJ, Westwood MJ. Computation of chemically reacting flow on parallel systems. Parallel Comput Fluid Dyn 1996;1995:113–20. Available from: https://doi.org/10.1016/B978-044482322-9/50068-2.

[25] Foley HC, Foley HC. Chapter 7 – Reacting systems—kinetics and batch reactors. In: Introd. to Chem. Eng. Anal. Using Math. 2002. 297–361. Available from: http://dx.doi.org/10.1016/B978-012261912-0/50009-X.

[26] Arnold K, Stewart M, Stewart MI, Stewart MI, Arnold K, Stewart M, et al. Chapter 8 – Pressure drop in piping. In: Surf. Prod. Oper. Des. Oil-Handling Syst. Facil. 1999. 244–84. Available from: http://dx.doi.org/10.1016/B978-088415821-9/50009-9.

[27] Celata GP. 5 – Pressure drop. In: Microchannel Phase Chang. Transp. Phenom. 2016. 193–216. Available from: http://dx.doi.org/10.1016/B978-0-12-804318-9.00005-4.

[28] Stewart M, Stewart M. 6 – Fluid flow and pressure drop. Surf. Prod. Oper. 2016. 343–470. Available from: http://dx.doi.org/10.1016/B978-1-85617-808-2.00006-7.

[29] Thomas D, Charvet A, Bardin-Monnier N, Appert-Collin J-C, Bardin-Monnier N, Thomas D. 3 – Initial pressure drop for fibrous media. In: Aerosol Filtr. 2017. 49–78. Available from: http://dx.doi.org/10.1016/B978-1-78548-215-1.50003-2.

[30] Park HJ, Tasaka Y, Oishi Y, Murai Y. Drag reduction promoted by repetitive bubble injection in turbulent channel flows. Int J Multiph Flow 2015;75:12–25. Available from: https://doi.org/10.1016/j.ijmultiphaseflow.2015.05.003.

[31] Fink, J. Chapter 6 – Drag reduction and flow improvement. In: Guid. to Pract. Use Chem. Refineries Pipelines; 2016. 83–108. Available from: http://dx.doi.org/10.1016/B978-0-12-805412-3.00006-4.

[32] Dey P, Das AK. Numerical analysis of drag and lift reduction of square cylinder. Eng Sci Technol Int J 2015;18:758–68. Available from: https://doi.org/10.1016/j.jestch.2015.05.007.

[33] Yang D, Xiong YL, Guo XF. Drag reduction of a rapid vehicle in supercavitating flow. Int J Nav Archit Ocean Eng 2017;9:35–44. Available from: https://doi.org/10.1016/j.ijnaoe.2016.07.003.

[34] Kundu PK, Cohen IM, Dowling DR, Kundu PK, Cohen IM, Dowling DR. Chapter 9 – Laminar flow. In: Fluid Mech. 2016. 409–67. Available from: http://dx.doi.org/10.1016/B978-0-12-405935-1.00009-5.

[35] Hirsch C, Hirsch C. Chapter 12 – Numerical solutions of viscous laminar flows. In: Numer. Comput. Intern. Extern. Flows. 2007. 599–XVI. Available from: http://dx.doi.org/10.1016/B978-075066594-0/50056-4.

[36] Churchill SW, Churchill SW. Chapter 14 – Laminar flow over wedges and disks. In: Viscous Flows. 1988. 301–15. Available from: http://dx.doi.org/10.1016/B978-0-409-95185-1.50024-3.

[37] Churchill SW, Churchill SW. Chapter 15 – Laminar flow over a circular cylinder. In: Viscous Flows. 1988. 317–57. Available from: http://dx.doi.org/10.1016/B978-0-409-95185-1.50025-5.

[38] Churchill SW, Churchill SW. Chapter 16 – Laminar flow over a solid sphere. In: Viscous Flows. 1988. 359–409. Available from: http://dx.doi.org/10.1016/B978-0-409-95185-1.50026-7.

[39] Churchill SW, Churchill SW. Chapter 11 – The Blasius solution for laminar flow along a flat plate. In: Viscous Flows. 1988. 255–69. Available from: http://dx.doi.org/10.1016/B978-0-409-95185-1.50021-8.

[40] Churchill SW, Churchill SW. Chapter 12 – Integral boundary-layer solution for laminar flow along a flat plate. In: Viscous Flows. 1988. 271–8. Available from: http://dx.doi.org/10.1016/B978-0-409-95185-1.50022-X.

[41] Ting DS-K, Ting DS-K. Chapter 1 – Introducing flow turbulence. In: Basics Eng. Turbul. 2016. 3–18. Available from: http://dx.doi.org/10.1016/B978-0-12-803970-0.00001-5.

[42] Venkatram A, Du S. TURBULENCE & MIXING | Turbulent diffusion. Encycl Atmos Sci 2015;277–86. Available from: https://doi.org/10.1016/B978-0-12-382225-3.00441-2.

[43] Gatski TB, Bonnet J-P, Gatski TB, Bonnet J-P. Chapter 3 – Compressible turbulent flow. In: Compressibility, Turbul. High Speed Flow. 2013. 39–77. Available from: http://dx.doi.org/10.1016/B978-0-12-397027-5.00003-4.

[44] Zienkiewicz OC, Taylor RL, Nithiarasu P, Zienkiewicz OC, Taylor RL, Nithiarasu P. Chapter 8 – Turbulent flows. In: Finite Elem. Method Fluid Dyn. 2014. 283–308. Available from: http://dx.doi.org/10.1016/B978-1-85617-635-4.00008-X.

[45] De Bortoli ÁL, Andreis GSL, Pereira FN, De Bortoli ÁL, Andreis GSL, Pereira FN. Chapter 4 – Mixing and turbulent flows. Model. Simul. React. Flows 2015;53–72. Available from: https://doi.org/10.1016/B978-0-12-802974-9.00004-0.

[46] Njobuenwu DO, Fairweather M. Large eddy simulation of non-spherical particle deposition in a vertical turbulent channel flow. Comput Aided Chem Eng 2014;33:907–12. Available from: https://doi.org/10.1016/B978-0-444-63456-6.50152-6.

[47] Cebeci T, Cebeci T. Chapter 2 – Conservation equations for compressible turbulent flows. In: Anal. Turbul. Flows with Comput. Programs. 2013. 33–51. Available from: http://dx.doi.org/10.1016/B978-0-08-098335-6.00002-1.

[48] Beljaars A. NUMERICAL MODELS | Parameterization of physical processes: turbulence and mixing. Encycl Atmos Sci 2015;200–11. Available from: https://doi.org/10.1016/B978-0-12-382225-3.00310-8.

[49] Meslem A, Bode F, Croitoru C, Nastase I. Comparison of turbulence models in simulating jet flow from a cross-shaped orifice. Eur J Mech – B/Fluids 2014;44:100–20. Available from: https://doi.org/10.1016/j.euromechflu.2013.11.006.

[50] Abdi DS, Bitsuamlak GT. Wind flow simulations on idealized and real complex terrain using various turbulence models. Adv Eng Softw 2014;75:30–41. Available from: https://doi.org/10.1016/j.advengsoft.2014.05.002.

[51] Vakamalla TR, Mangadoddy N. Numerical simulation of industrial hydrocyclones performance: role of turbulence modelling. Sep Purif Technol 2017;176:23–39. Available from: https://doi.org/10.1016/j.seppur.2016.11.049.

[52] Wilson DR, Craft TJ, Iacovides H. Application of Reynolds stress transport turbulence closure models to flows affected by Lorentz and buoyancy forces. Int J Heat Fluid Flow 2015;55:180–97. Available from: https://doi.org/10.1016/j.ijheatfluidflow.2015.06.007.

[53] Elkhoury M. Assessment of turbulence models for the simulation of turbulent flows past bluff bodies. J Wind Eng Ind Aerodyn 2016;154:10–20. Available from: https://doi.org/10.1016/j.jweia.2016.03.011.

[54] Robertson E, Choudhury V, Bhushan S, Walters DK. Validation of OpenFOAM numerical methods and turbulence models for incompressible bluff body flows. Comput Fluids 2015;123:122–45. Available from: https://doi.org/10.1016/j.compfluid.2015.09.010.

[55] Ting DS-K, Ting DS-K. Chapter 8 – Vortex dynamics. In: Basics Eng. Turbul. 2016. 165–86. Available from: http://dx.doi.org/10.1016/B978-0-12-803970-0.00008-8.

[56] Ting DS-K, Ting DS-K. Chapter 6 – Wall turbulence. In: Basics Eng. Turbul. 2016. 119–38. Available from: http://dx.doi.org/10.1016/B978-0-12-803970-0.00006-4.
[57] Chattopadhyay K, Guthrie RIL. Chapter 4.2 – Turbulence modeling and implementation. In: Treatise Process Metall. 2014. 445–451. Available from: http://dx.doi.org/10.1016/B978-0-08-096984-8.00008-2.
[58] Ting DS-K, Ting DS-K. Chapter 5 – Turbulence simulations and modeling. In: Basics Eng. Turbul. 2016. 99–118. Available from: http://dx.doi.org/10.1016/B978-0-12-803970-0.00005-2.
[59] Blazek J, Blazek J. Chapter 7 – Turbulence modeling. In: Comput. Fluid Dyn. Princ. Appl. 2015. 213–52. Available from: http://dx.doi.org/10.1016/B978-0-08-099995-1.00007-5.
[60] Cebeci T, Cebeci T. Chapter 6 – Transport-equation turbulence models. In: Anal. Turbul. Flows with Comput. Programs; 2013. 211–35. Available from: http://dx.doi.org/10.1016/B978-0-08-098335-6.00006-9.
[61] Karvinen A, Ahlstedt H. Comparison of turbulence models in case of jet in crossflow using commercial CFD code. Eng Turbul Model Exp 2005;6:399–408. Available from: https://doi.org/10.1016/B978-008044544-1/50038-8.
[62] Cebeci T, Cebeci T. Chapter 8 – Differential methods with algebraic turbulence models. In: Anal. Turbul. Flows with Comput. Programs; 2013. 293–356. Available from: http://dx.doi.org/10.1016/B978-0-08-098335-6.00008-2.
[63] Casey MV. Validation of turbulence models for turbomachinery flows – a review. Eng Turbul Model Exp 2002;5:43–57. Available from: https://doi.org/10.1016/B978-008044114-6/50005-3.
[64] Moeng C-H, Sullivan P. Large eddy simulation. Encycl Atmos Sci 2003;1140–50. Available from: https://doi.org/10.1016/B0-12-227090-8/00201-3.
[65] Ciofalo M. Large-eddy simulation: a critical survey of models and applications. Adv Heat Transf 1994;25:321–419. Available from: https://doi.org/10.1016/S0065-2717(08)70196-5.
[66] Tsubokura M, Kobayashi T, Taniguchi N. Development of the subgrid-scale models in large eddy simulation for the finite difference schemes. Eng Turbul Model Exp 2005;6:297–306. Available from: https://doi.org/10.1016/B978-008044544-1/50028-5.
[67] Moeng C-H, Sullivan PP. NUMERICAL MODELS | Large-eddy simulation. Encycl Atmos Sci 2015;232–40. Available from: https://doi.org/10.1016/B978-0-12-382225-3.00201-2.
[68] Bhushan S, Alam MF, Walters DK. Evaluation of hybrid RANS/LES models for prediction of flow around surface combatant and suboff geometries.

Comput Fluids 2013;88:834–49. Available from: https://doi.org/10.1016/j.compfluid.2013.07.020.

[69] Ashton N, West A, Lardeau S, Revell A. Assessment of RANS and DES methods for realistic automotive models. Comput Fluids 2016;128:1–15. Available from: https://doi.org/10.1016/j.compfluid.2016.01.008.

[70] Ruiz J, Kaiser AS, Zamora B, Cutillas CG, Lucas M. CFD analysis of drift eliminators using RANS and LES turbulent models. Appl Therm Eng 2016;105:979–87. Available from: https://doi.org/10.1016/j.applthermaleng.2016.01.108.

[71] Tallmadge JA. In: Collins RE, editor. Flow of Fluids Through Porous Materials. New York, NY: Reinhold Publishing Co; 1961270 pp. $12.50. AIChE J 1962;8:2–2. Available from: https://doi.org/10.1002/aic.690080102.

[72] Brown G. Darcy and the Pitot tube. Int. Eng. Hist. Herit. 2001.

[73] Urquiza JM, N'Dri D, Garon A, Delfour MC. Coupling Stokes and Darcy equations. Appl Numer Math 2008. Available from: https://doi.org/10.1016/j.apnum.2006.12.006.

[74] Hirani AN, Nakshatrala KB, Chaudhry JH. Numerical method for Darcy flow derived using discrete exterior calculus, Int. J. Comput. Methods Eng. Sci. Mech. 16(3), 151–169 (2015).

[75] Imomnazarov KK. Modified Darcy laws for conducting porous media. Math Comput Model 2004. Available from: https://doi.org/10.1016/j.mcm.2004.01.001.

[76] Kraft R, Yaakobi D. Some remarks on non-Darcy flow. J Hydrol 1966. Available from: https://doi.org/10.1016/0022-1694(66)90077-1.

[77] Vassilev D, Yotov I. Coupling Stokes-Darcy flow with transport. SIAM J Sci Comput 2009. Available from: https://doi.org/10.1137/080732146.

[78] Awad MM. An alternative form of the Darcy equation. Therm Sci 2014. Available from: https://doi.org/10.2298/TSCI131213042A.

[79] Liu X, Civan F, Evans RD. Correlation of the non-Darcy flow coefficient. J Can Pet Technol 1995. Available from: https://doi.org/10.2118/95-10-05.

[80] Wu Y-S, Wu Y-S. Chapter 8 – Non-Darcy flow of immiscible fluids. In: Multiph. Fluid Flow Porous Fract. Reserv. 2016. Available from: http://dx.doi.org/10.1016/B978-0-12-803848-2.00008-8.

[81] Brown G, Oklahoma SU. History of the Darcy-Weisbach equation for pipe flow resistance. Envior Water Resour Hist 2002. Available from: https://doi.org/10.1061/40650(2003)4.

[82] Badea L, Discacciati M, Quarteroni A. Numerical analysis of the Navier-Stokes/Darcy coupling. Numer Math 2010. Available from: https://doi.org/10.1007/s00211-009-0279-6.

[83] Wu YS. Numerical simulation of single-phase and multiphase non-Darcy flow. Transp Porous Media 2002. Available from: https://doi.org/10.1023/A:1016018020180.

[84] Darcy H. Darcy's law. Ground Water 1994. Available from: https://doi.org/10.1007/BF02120313.

[85] Barree RD, Conway MW. Beyond beta factors: a complete model for Darcy, Forchheimer, and Trans-Forchheimer flow in porous media. Spe 2004;89325. Available from: https://doi.org/10.2523/89325-MS.

[86] Peksen M. Numerical modelling of nonwoven thermal bonding process & machinery. PhD Thesis. UK: Loughborough University; 2008.

[87] Peksen M, Acar M, Malalasekera W. Computational modelling and experimental validation of the thermal fusion bonding process in porous fibrous media. Proc Inst Mech Eng Part E-J Process Mech Eng 2011. Available from: https://doi.org/10.1177/0954408910396785.

[88] Straughan B. Structure of the dependence of Darcy and Forchheimer coefficients on porosity. Int J Eng Sci 2010. Available from: https://doi.org/10.1016/j.ijengsci.2010.04.012.

[89] Ruth D, Ma H. On the derivation of the Forchheimer equation by means of the averaging theorem. Transp Porous Media 1992. Available from: https://doi.org/10.1007/BF01063962.

[90] Moutsopoulos KN, Tsihrintzis VA. Approximate analytical solutions of the Forchheimer equation. J Hydrol 2005. Available from: https://doi.org/10.1016/j.jhydrol.2004.11.014.

[91] Sidiropoulou MG, Moutsopoulos KN, Tsihrintzis VA. Determination of Forchheimer equation coefficients a and b. Hydrol Process 2007. Available from: https://doi.org/10.1002/hyp.6264.

[92] Whitaker S. The Forchheimer equation: a theoretical development. Transp Porous Media 1996. Available from: https://doi.org/10.1007/BF00141261.

[93] Chen Z, Lyons SL, Qin G. Derivation of the Forchheimer law via homogenization. Transp Porous Media 2001. Available from: https://doi.org/10.1023/A:1010749114251.

[94] Aulisa E, Bloshanskaya L, Efendiev Y, Ibragimov A. Upscaling of Forchheimer flows. Adv Water Resour 2014. Available from: https://doi.org/10.1016/j.advwatres.2014.04.016.

[95] Forchheimer P, Boden WD. Zeit. Ver. Deutsch. Ing. 45. Zeitschrift Des Vereins Dtsch Ingenieure 1901.

[96] Launder BE, Spalding DB. The numerical computation of turbulent flows. Comput Methods Appl Mech Eng 1974;3:269–89. Available from: https://doi.org/10.1016/0045-7825(74)90029-2.

[97] Capone DE, Lauchle GC. Modelling the unsteady axial forces on a finite-length circular cylinder in cross-flow. J Fluids Struct 2002;16:667–83. Available from: https://doi.org/10.1006/jfls.2002.0439.

[98] Lima E, Silva ALF, Silveira-Neto A, Damasceno JJR. Numerical simulation of two-dimensional flows over a circular cylinder using the immersed boundary method. J Comput Phys 2003;189:351–70. Available from: https://doi.org/10.1016/S0021-9991(03)00214-6.

## 2.4 Problems

The introduced problems aim to familiarise the analyst with basic fluid dynamics variables whilst improving the understanding and requirements of different meshing methods. The analyst can use the following instructions, where appropriate:

The imposed inlet velocity should be calculated from:

$$Re = \frac{\rho U_\infty D}{\mu}$$

such that the desired Reynolds number ($Re$) is achieved. $D$ is the cylinder diameter. The fluid properties $\rho$ and $\mu$ are considered to be constant. When the velocity and pressure fields are calculated, the drag and lift coefficients, as well as the Strouhal number can also be determined. This can be performed using directly the force field or through the following relations:

$$C_D = \frac{F_D}{(1/2)\rho U_\infty^2 D}$$

$$F_D = -\int_0^L f_x ds$$

The drag coefficient $C_D$ with the drag force $F_D$ and the $x$ component of the Lagrangian force $f_x$ can be obtained in most software as an output. $L$ refers to the length of the interface inside the domain. Likewise, the lift coefficient $C_L$ can be calculated using the $y$ component of the Lagrangian force expressed as:

$$C_L = \frac{F_L}{(1/2)\rho U_\infty^2 D}$$

$$F_{Li} - \int_0^L f_y ds$$

Finally, the analyst can determine the Strouhal number that refers to the dimensionless frequency with which the vortices are shed behind the obstacle. It can be expressed as:

$$St = \frac{fD}{U_\infty}$$

where $f$ is the vortex shedding frequency. This frequency can be obtained using the lift coefficient time distribution that provides the vortices period.

**2.1** The geometry and dimensions depicted in Fig. 2.P1 are considered for a fluid flow analysis. A cylinder is placed inside a rectangular fluid domain. The distances from the inlet to the centre of the cylinder, as well as to the outlet region are specified. Calculate the given tasks, using a Reynolds number of $Re = 100$ and water as the fluid. Perform a transient analysis for 200 s and determine the following:

**a.** Show the velocity distribution of the fluid domain for instants of 50, 100 and 200 s.

**b.** Show the pressure components in $x$- and $y$-direction.

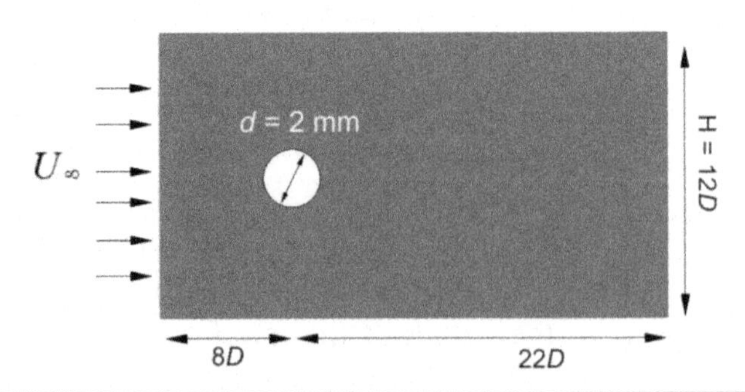


**Figure 2.P1** Problem description.

**c.** Calculate the drag and lift coefficients.
**d.** Calculate the Strouhal number.
**e.** Perform the same analysis, considering a steady state and observe the velocity distribution.

**2.2** A rectangular air domain is used to simulate the flow over a stationary cylinder. The boundary conditions are imposed in such a way that the flow is from the left to the right, as depicted in Fig. 2.P2. A circular cylinder is placed inside the domain centrally. When a Reynolds number of $Re = 80$ is used, perform a steady state analysis for the following tasks:

**a.** Use a uniform grid and perform a study using three different grid sizes to verify the results are independent of the used grid, compare the results using the drag coefficient $C_D$.
**b.** Show the pressure and velocity field in $x$- and $y$-direction.
**c.** Perform the same analysis, considering a Reynolds number of $Re = 100$ and $Re = 60$. Compare the differences in the lift coefficients.

**2.3** Water flows in a two-dimensional domain as shown in Fig. 2.P3. To simulate the fluid flow, two same sized stationary cylinders are placed inside the domain. The inlet flow is uniform and a Reynolds number of $Re = 150$ is considered. Perform a steady state analysis for the solution of the following tasks:

**a.** Use a multiblock grid to resolve the region of the cylinders and calculate the velocity in $x$-direction.
**b.** Determine the pressure drop inside the domain.

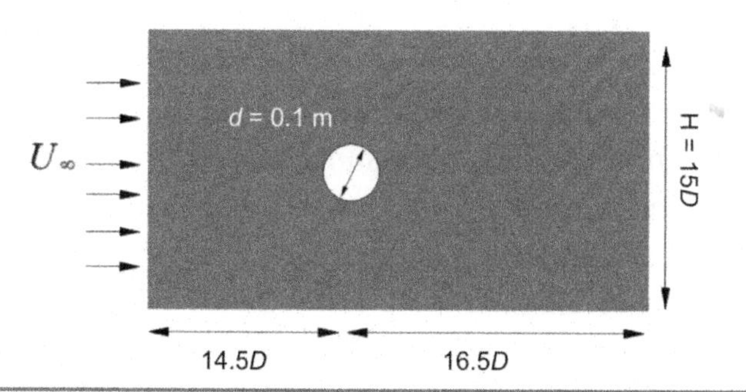


**Figure 2.P2** Problem description.

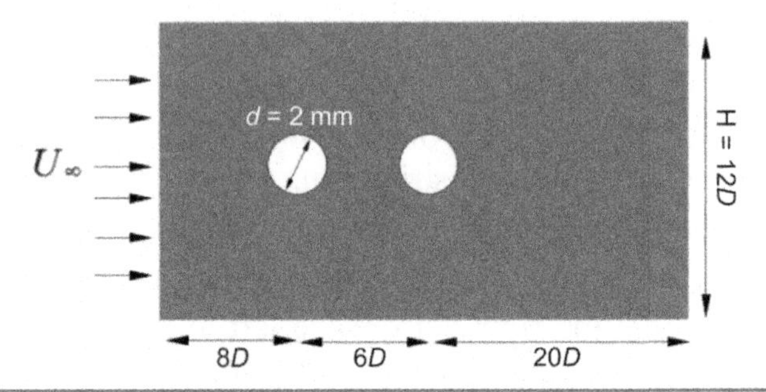


**Figure 2.P3** Problem description.

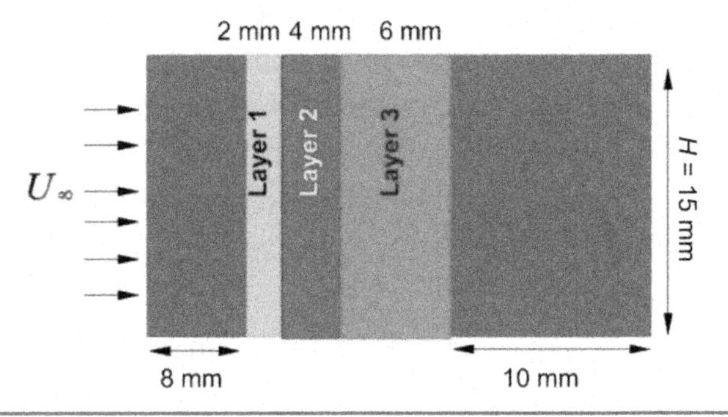


**Figure 2.P4** Problem description.

**c.** Perform the same analysis, considering a Reynolds number of $Re = 100$ and $Re = 60$. Compare the differences in the lift coefficients.

**d.** Calculate the problem using air as fluid medium for $Re = 150$ and compare your drag coefficient result with the drag coefficient determined using water as fluid medium.

**2.4** Air flows in a two-dimensional box as shown in Fig. 2.P4. A filter material made of three different layers is placed inside the domain. The materials are assumed as porous. The inlet flow is uniform and a velocity of 1 m/s is considered. The materials resist to the fluid flow in the flow direction with properties given in Table 2.P1. The top and bottom sides of the solid box can be considered as walls. The outlet region of the domain is

**TABLE 2.P1** Material Properties Used for the Porous Media

| Material | Porosity | Viscous Resistance ($1/m^2$) for Both Directions |
|---|---|---|
| Layer 1 | 0.9 | 819,850 |
| Layer 2 | 0.7 | 795,208 |
| Layer 3 | 0.4 | 2.0 $E^{+08}$ |

set to outflow boundary conditions. Perform a steady state analysis for the solution of the problem.

**a.** Determine the pressure drop.

**b.** Calculate the velocities at the inlet and outlet regions of each layer.

# Multiphysics Modelling of Thermal Environments

## Chapter Outline



Modelling of thermal environments is a specialised subdiscipline of multiphysics that encompasses exclusively the thermal energy and its heat exchange between not only different media, but also into other usable forms of energy. The so-called computational heat transfer [1–6] discipline is armed with the expertise to design systems and processes to convert generated energy from various thermal sources into chemical, mechanical or electrical energy, depending on the particular problem at hand. This affinity to interactions with various disciplines emphasises its importance. Obviously, knowledge and expertise of all aspects of



Multiphysics Modelling. DOI: https://doi.org/10.1016/B978-0-12-811824-5.00003-1

heat transfer modes [7–11] are required to handle this kind of challenging engineering problems.

In science and engineering, the transformation of energy within different forms occurs due to an exchange of energy between the systems such as chemical energy and kinetic energy. When this exchange is based on energy transfer without any exchange of mass or no temperature difference is present, it is said to have been an exchange based on work. On the other hand, if the energy exchange occurs due to a temperature difference, it is said to be transferred by heat. It is the heat transfer and its physical laws governing its exchange, which is the subject of this chapter.

It should be noted that an exchange of energy by heat between two systems may also be different sections within the same body and that this exchange refers to the transfer of internal energy of its thermofluid, the enthalpy between this sections. The second law of thermodynamics [12–14] requires that the heat transfer between two medium takes place from the hotter system to the colder one.

Like fluid dynamics, almost in any scientific and engineering problem, an exchange of energy due to heat has been encountered. Today's global heating problem is perhaps the most challenging application of heat transfer. Even the modern alternative and regenerative energy sources, as well as other means of power sources, involve the production of work from various means of boilers, turbines, heat exchangers, etc. All these equipment involve a transfer of heat [15–24]. The understanding of highly complex combustion processes [25–30], furnace operation [31,32] or cooling processes, etc. Even in building technologies where internal temperature changes are tried to be reduced, require a thorough knowledge of various heat transfer interactions.

Modern computer CPUs, temperature sensors, nanotechnologies such as nanotubes, as well as biomedical applications such as thermograph, and food technologies like pasteurising comprise heat transfer processes [33,34]. Thus the demand to understand and quantify such various technical and natural problems has become increasingly important, as high-tech engineering has become more complex. This complexity requires immense knowledge of heat transfer as to calculate, analyse and optimise thermal processes and to develop new technologies.

Generally, the processes in thermal environments and heat transfer underlie some mechanisms that needs to be understood prior modelling them. Giving some foundations, shedding light on some questions such as how the heat transfer is realised and the correct use of some particular terms would be quite beneficial. Beginning from the latter one, from the study of thermodynamic courses it should be well recalled that in a heat exchange basically two systems are considered. When an exchange between the surroundings and the system is not allowed, no heat is gained or lost then we speak about *adiabatic systems.* Otherwise, the heat permeable *diathermic systems.* As heat transfer obviously deals with heat, it is important to comprehend the terms *temperature*, *temperature field* and *temperature gradient*, as the correct use will be important in developing user-defined macros and in the postprocessing of numerically predicted results.

Temperature is from a thermodynamic point of view an intensive state variable. This means, it is unlike an extensive state variable, independent of the mass of the system. It is a state variable because it is describing the thermodynamic system independent of the path of a process. Temperature must be defined and cannot be extracted from other variables. Thus it is characterising the thermal state of a system.

Heat or work have been describing quantitatively the transition route between the equilibrium states and depend on the path, thus are referred as process variables. Worth to mention is to take care in the use of the absolute temperature Kelvin (K) and Celsius (°C) because when using temperature distributions and profiles in programming activities, the numerical code will usually consider the input quantities as Kelvin, thus misleading results and errors may arise. Temperature is a scalar variable, i.e. is independent of direction, however, temperature field is dependent on space and time, thus is a scalar field (three-dimensional space). In practical steady state situations the time effect is usually not considered.

The temperature gradient is not a scalar, it is a vector. The misuse of the term is very often seen in analysing heat transfer results. It can be defined as the change in temperature over the change in distance, or at what rate the temperature changes the most rapidly in a direction. It is derived from the temperature field according the space coordinates. Hence, it is described as:

$$\text{grad}T = \nabla T = \left(\frac{\partial T}{\partial x}, \frac{\partial T}{\partial y}, \frac{\partial T}{\partial z}\right) \tag{3.1}$$

An example to simplify the understanding would be to consider that the temperatures of given points A and B are 10°C and 20°C, respectively. The distance in $x$-direction between them is 100 km. The temperature gradient would then be:

$$\frac{\partial T}{\partial x} = 10^{\circ}\text{C}/100\ \text{km} = 0.1^{\circ}\text{C}/\text{km}$$

As a multiphysics analyst it is important to understand the processes, which underlie the heat transfer mechanisms, thus let's proceed with the modes of heat transfer and the governing equations used in modelling thermal environments. Generally speaking, from the fundamental studies of heat transfer, it is well known that heat is being transferred from a warmer zone inside a medium or between media through some mechanisms. Having defined the temperature gradient, it can be confidently said that when a temperature gradient within a solid or a stagnant fluid medium exists, technically the first mode of heat transfer, i.e. heat conduction is concerned.

The temperature gradients and the associated energy diffusion/transfer result due to atomic and molecular activities. Recall that the heat conduction is governed by the Fourier's law that most of the engineers are familiar of simplified conditions. However, in complicated multiphysics analyses, the multidimensional form of the law will be used to determine the heat flux.

The second mechanism called the heat convection is of concern, as soon as an energy transfer between a surface and a fluid in motion occurs (in the presence of a temperature difference). The heat exchange is dominated by the bulk motion of the fluid particles. Therefore the reader should value its importance and the association to fluid flow presented in Chapter 2, Multiphysics Modelling of Fluid Flow Systems, where the energy aspects were on purpose not covered. Thus the heat convection is important to understand and calculate the energy carrying fluid flow problems (usually referred to as thermo-fluid) in external or internal flows, as well as phase change problems such as melting and solidification or boiling, condensation [35–37].

The fluid motion may be caused by some source of external means such as a fan, in which the process is called *forced convection*. However, if the fluid motion is caused due to density differences that are caused by temperature differences in the fluid mass, then the process is termed as *free or natural convection* [35]. Virtually, it is not possible to observe heat conduction solely in a fluid, because as soon as temperature differences arise in a fluid, natural convection currents will occur as a result of the density differences. The fundamental laws of heat conduction must be coupled with the fluid motion equations to describe correctly the heat convection, thus Newton's law of cooling lumped these together, which the reader is aware of from fundamental courses.

The last fundamental mechanism the heat is transferred, is the heat radiation which has very important industrial applications such as heating, cooling, drying and energy conversion and solar radiation [38–41]. The most distinguishing feature compared to the former two is that the heat radiation does not require some form of matter such it was the case for the previous ones in form of temperature gradients. The heat radiation is based on the emission mechanism, which results from the oscillation and the transmission of electrons that build up matter.

Usually it is considered as the propagation of electromagnetic waves. These movements are sustained by internal energy and the associated temperature of the matter. All forms of matter emit radiation. For some gases and semitransparent solids it is a volumetric process, whereas for most solids and liquids it is considered to be a surface phenomenon [42].

## 3.1 Governing Equations

The governing equations that describe the heat transfer in thermal environments are categorised into three basic types or usually referred as modes. Although it will be apparent that as one is involved in multiphysics, it is certainly a rare situation when an analyst encounters a problem of practical importance, which does not include at least two, if not all of them interacting simultaneously. As soon as heat transfer is considering convection, it is necessary to determine the fluid temperature at each point within the flow field. This dependent variable requires the use of an additional fundamental equation, in addition to the governing equations of continuity and momentum. This additional relation known as the energy equation is a statement of the derivation of the first law of thermodynamics.

This obviously comprises energy, heat transfer and work. If we consider the general control volume formula for the first law of thermodynamics

$$\frac{\partial \rho}{\partial t}\int_{Cv} e\rho d\mathcal{V} + \int_{S(t)} e\rho U.ndA = \left(\dot{Q}_{net(in)} + \dot{W}_{net(in)}\right)_{Cv} \tag{3.2}$$

The term $e$ represents the total stored energy per unit mass for a fluid entering, leaving and within the control volume. It is related to the internal energy per unit mass, the kinetic energy per unit mass and the potential energy per unit mass. The term that expresses the time rate of change of the total stored energy of the contents of the control volume vanishes for the steady state situation. (Recall that the equation is the general Reynolds transport equation, involving the volume and surface integrals with $e$, i.e. the total energy stored per unit mass given as the property changing in the control volume.)

It is the term $\dot{Q}$ representing any means of exchange of energy between the environment and the control volume due to a difference in temperature. Therefore all modes of heat transfer are covered. From the integral form of scalar quantities it is well known that the general form of the control volume equation on the left side will only be substituted with the appropriate scalar quantity, which in the heat transfer case it has been referred to enthalpy.

$$\frac{\partial \rho}{\partial t}\int_{Cv} h\rho d\forall + \int_{S(t)} h\rho U.ndA = \int_{S(t)} k\nabla T.ndA + \int_{Cv} (u.\nabla p + \tau{:}\nabla u)d\forall + \frac{\partial \rho}{\partial t}\int_{Cv} Sd\forall \tag{3.3}$$

where $h$ is the enthalpy, $T$ the temperature, $u$ the fluid velocity, $k$ the thermal conductivity and $\tau$ the viscous part of the stress tensor. $S$ are the source terms due to any additional means such as Joule effects due to Ohmic resistance, the electrochemical reactions or the heat radiation, according to the domain of interest.

The first term represents the time rate of change of the enthalpy of the control volume; the second term on the left hand side describes the transport due to convection. The right hand side depicts the heat conduction, work due to pressure forces and viscous dissipation. Another often used description, especially when incompressible effects are negligible, is the coordinate free vector form such as:

$$\frac{\partial \rho h}{\partial t} = -\nabla(\rho h u) - \nabla \dot{q} - (\tau{:}\nabla u) - p(\nabla.u) + S \tag{3.4}$$

(Notice that in this form, it has been assumed that the enthalpy equals directly to internal energy, thus the kinetic energy has been omitted, which the enthalpy $h$ normally comprises as well.)

On the other hand, when describing the relation between the flow eld and the heat transfer, density changes arise as mentioned earlier. These have been referred to the pressure and temperature changes. These variables need to be linked together. The equation of state, comprising the temperature and density as state variables can achieve this through the use of state equations of pressure and specific internal energy ($ie$):

$$\begin{aligned} p &= p(\rho, T) \\ ie &= ie(\rho, T) \end{aligned} \tag{3.5}$$

$$\begin{aligned} p &= \rho RT \\ ie &= CT \end{aligned}$$

where $R$ refers to the ideal gas constant and $C$ is the specific heat. Notice that without density changes there would be no need for this link and use of energy equation. The energy equation is only solved when heat transfer is present. Recall that within the general form of the energy equation, the major changes occur within the $\dot{Q}$ term and the source term. Thus adapting to certain processes will result in the inclusion of new variables to the energy equation. Examples would be for instance, if combustion processes are accounted for, a common expressed form becomes:

$$\frac{\partial \rho h}{\partial t} + \nabla(\rho h u) = \nabla.\left(\frac{k}{c_p}:\nabla h\right) + S \tag{3.6}$$

where species diffusion and heat conduction terms are combined ($k$ for thermal conductivity and $cp$ the specific heat capacity). The total enthalpy is defined as:

$$h = \sum_j X_j h_j \tag{3.7}$$

with $X_j$ the mass fraction of the species $j$ and

$$h_j = \int_{T_{refj}}^{T} cp_j dT + h_j^0(Tref_j) \tag{3.8}$$

accounting to the formation of sensible enthalpy of the species $j$ at the specified reference temperature.

Regardless of combustion, when species transport problems are considered, the term

$$h = \nabla.\left(\sum_j X_j \vec{h}_j\right) \tag{3.9}$$

is added to the energy equation. When chemically reacting species transport would be included, then the energy due to chemical reaction would be included as a source term $S$ to the energy equation (recall that it was combined in the combustion example) expressed as:

$$S_{cr} = -\sum_j \frac{h_j^0}{M_j} R_j \tag{3.10}$$

The volumetric rate of creation of the species $R_j$ would appear in the equation. Processes like radiation or interphase sources, such as in spray, particles are added as source terms. Ultimately, when the heat transfer is calculated in the solid regions, the energy equation can be expressed in the form such as:

$$\frac{\partial \rho h}{\partial t} + \nabla(\rho h u) = \nabla.(k \nabla T) + S \tag{3.11}$$

the convective term on the left hand side is mainly due to the motion of the solid. The velocity field is computed from the motion specified to the solid zones. The heat flux and if any volumetric heat sources are present in the solid, they are again the terms appearing on the right side of the equation.

Where $h$ will be calculated from:

$$h_j = \int_{T_{refj}}^{T} cp_j dT + h_j^0 \tag{3.12}$$

Having elucidated the general and some special forms of the energy equations for heat transport problems, a special form of the convection is very often neglected. The natural convection is a form of heat transport that many applications are subjected to. As fluid warms, temperature differences among regions become present, and therefore rise through the action of buoyancy is observed.

Flow is induced by force of gravity acting on the density variation. Note that in the absence of this movement, the heat transfer would be due to heat conduction and its rate would be much lower. The underlying term is the buoyancy force that under gravitational circumstances acts as a force, which pushes a lighter region inside a heavier fluid upward leading to a rising fluid motion. Fluid for the buoyant force in the momentum equations is the difference in density, which will be locally different compared to the reference temperature.

The dimensional Rayleigh number (*Ra*) takes the role of the Reynolds number to determine the effect of the natural convection. The effect will be more dominant with large variations in density between the fluids, or when a larger distance through the convecting medium such as depth is present. In cases where convection occurs due to heating, simplifications are performed and the so called Boussinesq approximation has been used to account for the buoyant effect through the expression:

$$(\rho - \rho_0)g = \rho_0 \alpha (T - T_0)g \tag{3.13}$$

In this assumption, instead of the density, the thermal expansion coefficient alpha of the fluid is used to compute the buoyant force.

The last mode of the heat transfer as elucidated is the heat radiation. The discussion of the previous so called modes has been limited to cases where a physical medium was required to transport the energy from the high temperature region to the low temperature sink region. The rate of thermal energy was proportional to the temperature difference between those regions. In practice, even if a body is insulated from the cooler regions, its temperature still decreases in time, showing loss of energy. In this case, a totally different mechanism takes place, which is the thermal radiation.

To exhibit the loss of energy, the body does not need to be solely heated. The emission of energy may be caused also due to oscillation, electrical current or even chemical reactions or electrochemical activity, etc. Moreover, this energy faces a body and is absorbed, it may manifest itself in the form of internal energy or as chemical reaction, force, etc. depending on the nature of the incident radiation and of course on the material properties of the body it strikes.

It should be noted and is strongly advised to go through all basic definitions such as emissive power, radiosity and irradiation. The concepts of the well-known black body, grey surfaces whilst intensively dealing within the multiphysics analyses need to be understood. Not only is the understanding of the process important, but also the material behaviour and properties of emissivity that will be required whilst predicting the thermal radiation.

For the scientists who predict the solar phenomena, it is strongly advised to improve their knowledge in the relation of emission and wavelength, because the fractions of wavelength of above and below 4 μm are particularly interesting. Temperatures of around 300K emit almost all of their energy portions at wavelengths less than 4 μm, but the Sun emits all its energy at the lower spectrum thus under 4 μm.

Hence, cooler bodies that are subjected to solar irradiation will emit different spectral characteristics than that incident upon it. This has an effect on the emissive and solar absorptive properties of real bodies. Another very important aspect to consider is the presence of absorbing and emitting gases among solid regions, where the gas phases must also be treated. Elementary gases with classical symmetrical molecules are transparent to thermal radiation, whereas gases that comprise complex molecules can emit and absorb within specific wavelengths thermal radiation. Water vapour or carbon dioxide is a typical example, which needs special attention. This is the reason that the book cannot cover all individual situations but draws attention to particular cases where the analyst should be careful. Thus to focus and reading on individual applications and carefully deciding comes prior starting the modelling of the real physical problem.

Concerning the modelling of thermal radiation, it has been obvious from the governing equations that it is treated as a separate equation as a source term. Most of the readers are familiar with its basic equation from the basic courses and its fundamentals based on Stefan–Boltzmann. The heat transfer due to radiation deserves special attention, as its value can easily rise to greater magnitudes than that of the convective and conduction heat transfer rates. However, the radiation is practically a very complex phenomenon and great care is required. Especially, due to its forth power exponents, it becomes predominant in high temperature applications.

Important applications include and are not limited to furnace operations, glass and ceramic processing, as well as heating ventilation and air conditioning. The general equation of the heat transfer due to radiation in a medium at a position $\vec{r}$ in the direction of $\vec{s}$ can be expressed as:

$$\frac{dI(\vec{r},\vec{s})}{ds} + \underbrace{(a+\sigma_s)I(\vec{r},\vec{s})}_{\text{Absorption}} = \underbrace{an^2\frac{\sigma T^4}{\pi}}_{\text{Emission}} + \underbrace{\frac{\sigma_s}{4\pi}\int_0^{4\pi} I(\vec{r},\vec{s_c})\Phi(\vec{s}\cdot\vec{s_c})d\Omega'}_{\text{Scattering}} \tag{3.14}$$

where the lower brace of absorption comprises radiation intensity as a function of the position and direction vectors $\vec{r},\vec{s}$, respectively. The absorption coefficient and the scattering coefficient are given as $a, \sigma_s$, respectively. The emission section includes the Stefan–Boltzmann constant $\sigma$ ($5.672 \times 10^{-8}$ W/m-K$^4$), which most of the readers are aware of the typical analytical equation; the absorption coefficient $a$ and the refractive index $n$. The last section of the radiation equation accounts for the scattering effects and includes the terms phase function ($\Phi$); solid angle $\Omega'$, as well as again the radiation intensity. The letters $s$ and $T$ stand for path length and local temperature, respectively.

The heat radiation equation predicts the local intensity of using the local absorption, out and in scattering, as well as emission properties of the media. It is affected in both directionally and spatiality. The fluid absorption is coupled with the energy equation and the radiation intensity is solved according several developed models. The following demonstrates one of the favourite heat radiation models, i.e. the discrete ordinate model [43,31]. The chosen radiation model is widely used in industrial applications applicable to the entire range of optical thicknesses, by comprising participating media, scattering, semitransparency (glass), specular surfaces such as mirrors, as well as wavelength-dependent transmission properties.

Thus limitation often seen in different models such as assumptions that all surfaces are diffuse is not the case. Therefore it is for complex multiphysics analyses despite its computational intensive manner, very appropriate. The DO model utilises the general radiation equation in a transformed form and uses it in the direction $\vec{s}$ as a field equation, thus it is expressed as:

$$\nabla \cdot (I(\vec{r},\vec{s})\vec{s}) + (a+\sigma_s)I(\vec{r},\vec{s}) = an^2\frac{\sigma T^4}{\pi} + \frac{\sigma_s}{4\pi}\int_0^{4\pi} I(\vec{r},\vec{s_c})\Phi(\vec{s}\cdot\vec{s_c})d\Omega' \tag{3.15}$$

One of the reasons the DO model has been introduced is its capability to predict solar load. Therefore it will be of interest to multiphysics analysts calculating radiative heating on automotive cabins and buildings, for example. The clue is that most of the solar ray tracing models do not calculate the emission from surfaces, thus the reflection of the incident load is assumed to be distributed uniformly across all participating surfaces rather than retained locally at the surfaces reflected to.

However, the discrete ordinates method is capable to calculate the radiation heat fluxes on semitransparent walls. This capability can be used to calculate using solar calculators the effective radiation load based on the position on the Earth's surface (latitude and longitude), the time of day, the season and established conditions as well as for clear or cloudy weather.

## 3.2 Practical Review Examples

Within the practical review section, it has been intended to introduce some examples of the numerically solved heat transfer problems. As heat transfer is one of the core areas multiphysics is dealing with, the reader should give particular attention to understand its theory and notice the wide range of applications within industrial and scientific applications. Thus it has been tried to give a flavour of solving the heat transfer modes numerically and to encourage the reader to implement the heat transfer modes in their complex problems.

The following example demonstrates a challenging conjugate heat transfer problem. The complex thermofluid flow comprises rotating flow domains together with porous media layers [44–47].

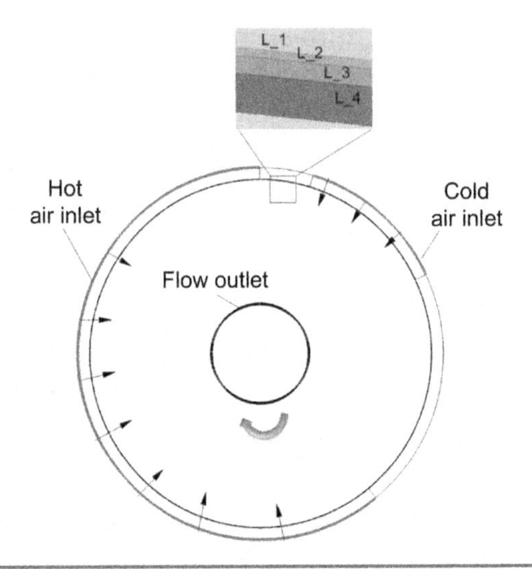


**Figure 3.1** Schematic description of the complex rotating thermofluid flow problem.

A rotating disc has been subjected to hot and cold air through two different inlet regions. A central hole is considered as the outflow area (Fig. 3.1). The disc comprises four layers that are considered to be as porous, where their properties will not be detailed.

Each porous fluid layer is separated by wall boundaries from each other. All components of the system are considered to be rotating at an angular velocity of $-0.68$ rad/s (clockwise). Thus a moving reference frame has been used. Turbulent flow has been assumed using a standard $k$-$\varepsilon$ model [48–51]. The hot flow stream passes through the porous layers and leaves the domain through the centrally defined outlet hole. During the rotation, the layers subjected to hot air flow are then subjected to the cold stream.

Fig. 3.2 depicts a screenshot of the radial velocity vector field in superposition with the translucent illustrated velocity magnitude contour plots. Visible is the angular movement of the velocity vectors due to the moving frame. The analysis resulted in a maximum velocity magnitude of around 11 m/s where the radial velocity has been predicted as $-7.25$ m/s.

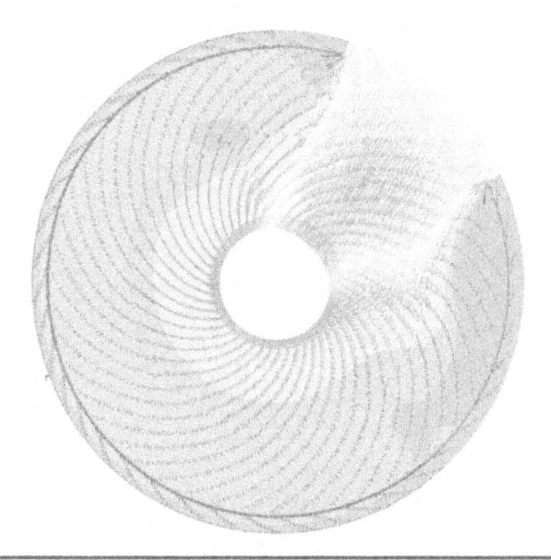

**Figure 3.2** Radial velocity vector field with velocity magnitude contours illustrated as superposition image.

As heat is carried by the fluid medium, i.e. forced convection takes place, the thermofluid flow behaviour over time has been simulated. A sequence of 20 s shows the development of the heat transferred from the hot air medium to the porous layers, having each different thermal properties and flow resistances. As expected, the outer layers are heated up faster compared to the inner regions, requiring approximately 20 s to reach a temperature close to the hot air inlet at the hot air inlet vicinity. The initial temperatures of the porous layers have been assumed at room temperature, explaining the sudden increase in temperature of the mid region (Fig. 3.3).

As the hot air flows through the cold layers, they absorb energy, thus the layers are slowly heated up. Meanwhile, the layers rotate with the system and are subjected to the cold air that flows through the second inlet. Thus the heat transferred to the porous layers is removed by the cool air. Hence, large temperature differences arise over time within the system. This kind of example will be a challenging task for the analyst; however, the multiphysics involved are widespread in industrial applications [35,47,52–57].

In the next example, a bundle of circular tubes is depicted, referring to a 2D view of distributed fibres. The cross section of the

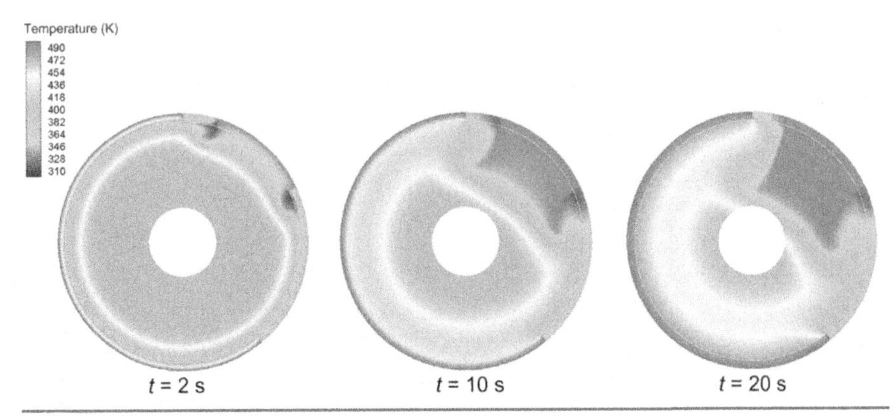


**Figure 3.3** Transient temperature distribution of the rotating disc over 20 s.

fibres shows that each circular cross section is made of two different materials. Convective hot air is assumed to pass through the fibres from the top edge. The surface temperatures are initially considered at room temperature. The air flow has a temperature and velocity of 500K and 0.0015 m/s, respectively.

The materials have thermal conductivities of $k_A = 0.28$ W/m-K and $k_B = 0.21$ W/m-K, densities of $\rho_A = 1310$ kg/m$^3$ and $\rho_B = 1320$ kg/m$^3$. The specific heat capacities are assumed as $C_{pA} = 1930$ J/kg-K and $C_{pB} = 1200$ J/kg-K. Fig. 3.4 gives an insight to the transient heating behaviour and the resulting temperature distribution of the fibres at an earlier stage. Remarkable are the thermal gradients, due to the material differences visible in some of the fibres. It should be noted that the thermofluid flow occurs around the fibres cross sections, thus the heat between the materials A and B results due to pure heat conduction, as they are considered as solid. It has been possible to interpret the heat transfer coefficients.

For example, at time instant of 1.4 s, a maximum surface heat transfer coefficient of 1773 W/m$^2$-K could be predicted at the fibre surface. The net total heat transfer rate of the analysis has been predicted as 63 W. It should be noted that due to the temperature differences of the free stream and the surfaces of the fibre cross sections, a thermal boundary layer is present. To observe this kind of effects, the numerical grid that surrounds each fibre cross section

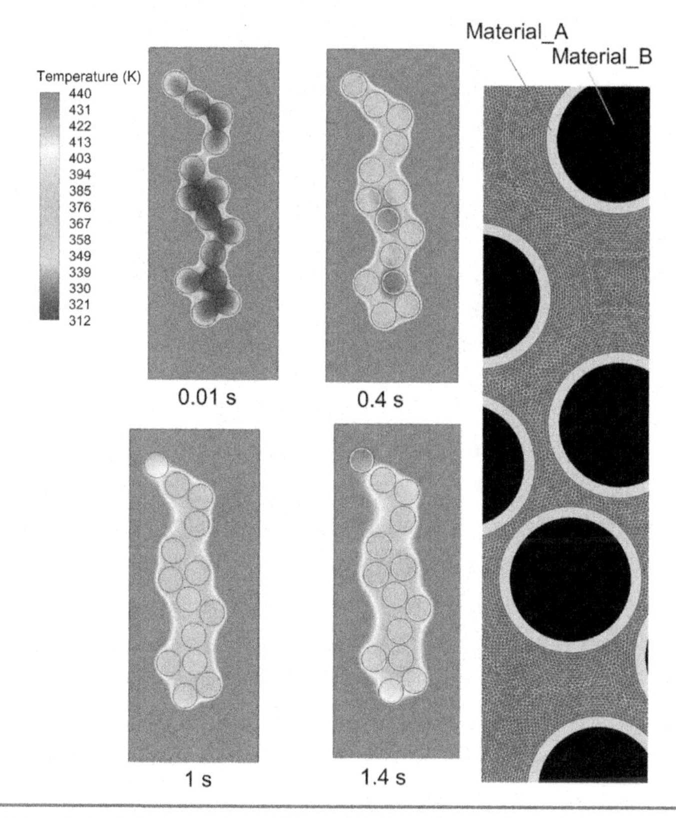


**Figure 3.4** Model domain with grid details and temperature distribution over four time instants.

should be resolved, such as it was demonstrated in Chapter 2, Multiphysics Modelling of Fluid Flow Systems.

High temperature furnaces are one of the most challenging multiphysics problems, involving multimodes of heat transfer. The following example demonstrates a metal component that has been heated within a conventional hood-type furnace. Heating proceeds from the surrounding walls of the hood as depicted in Fig. 3.5. The walls are heated using a temperature profile, enabling a stepwise increase, according to a ramped function. The air inside the furnace domain is also considered. Thus the buoyancy effects, arising due to density change, have been considered using gravitational acceleration. The air inside the furnace has been considered as transparent, whereas the metal component is assumed to be an absorbing,

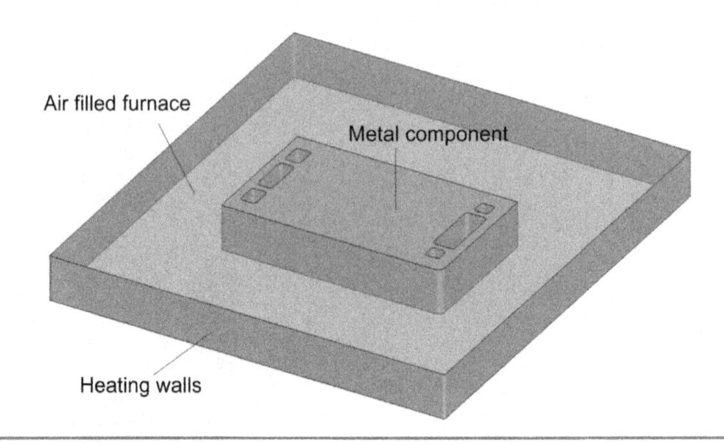


**Figure 3.5** Schematic description of the problem.

emitting nonscattering grey media. The absorption coefficient is assumed to be the same as the emitting coefficient and is distributed isotropically within the material (assumed as 0.3).

The predicted transient thermal field has been illustrated in Fig. 3.6. Due to the heat radiation that has been emitted from the surrounding walls, the heat is transferred through the air domain and partially absorbed by the metal component. It should be noted that the cavities inside the component are also filled with air, thus considered as participating media. Due to the density changes, natural convection occurs, causing a predicted slight motion of 0.03 m/s. The temperature increase over time shows in detail how the component interior still retains cooler zones compared to the edges and the exterior borders.

Each time frame has utilised its own legend, as to visualise the different thermal zones inside the domain. The analyst should pay attention to utilise exact dimensions whilst modelling thermal radiation problems. It is well known from fundamental courses that the distances between surfaces are important for the calculation of the thermal radiation, regardless of using view factors or discrete solid angles. The thermal radiation is an important mode, as well as the consideration of natural convection that is very often neglected in complex processes. The reader can easily realise its power, where optimisation of heating processes, as well as the precise programming of thermal controllers are of concern.

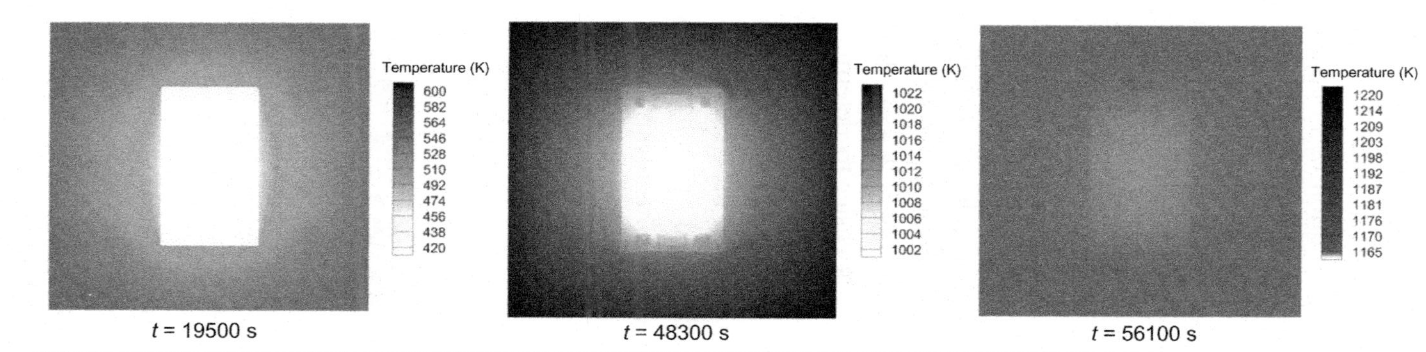


**Figure 3.6** Transient temperature distribution inside the furnace.

The last example has been chosen for a standalone heat transfer mode, i.e. heat conduction. A geometrically more complex sample has been chosen, as the sole solution of Fourier's law of heat conduction is pretty simple to be applicable on simpler geometries. A rim component of a car is subjected to different surface temperatures to evaluate its thermal behaviour. A symmetrical half model has been employed for the analysis. The boundary conditions together with the resulting temperature distribution are as illustrated in Fig. 3.7. The rim is assumed to be made of steel.

Constant surface temperature boundary conditions are applied, i.e. the surfaces maintain their fixed temperatures. As heat is transferred in the direction from hot to cold, the temperature between the cold and hot zones will change until a steady state has been achieved and no more energy exchange takes place. Recall that once the temperature distribution is known, it is easier to determine the heat conduction rate or the heat flux on any surface. For example, the surface where the 600°C has been applied predicts an elemental average heat flux of 0.11 W/mm$^2$ for a surface area of 4092 mm$^2$. This kind of information is often utilised by non multiphysics specialists as boundary condition for thermomechanical analyses.

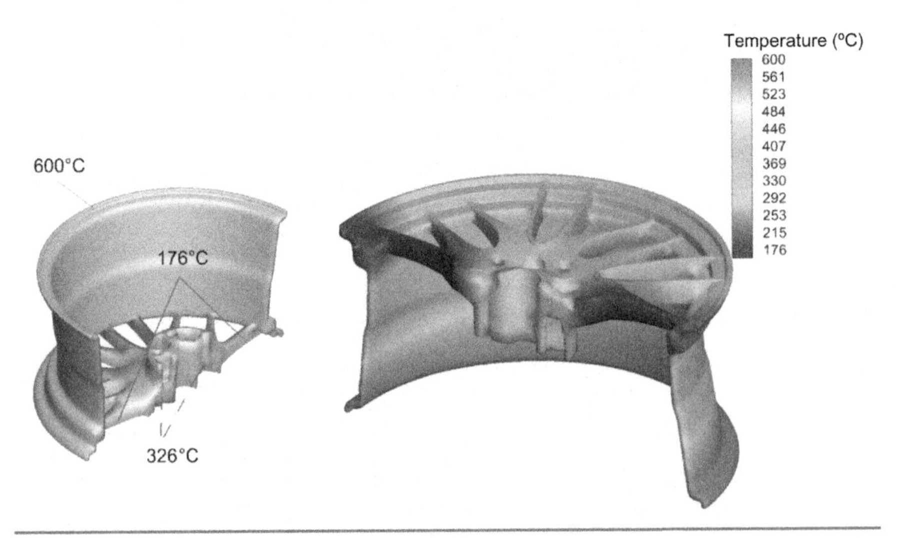


**Figure 3.7** Rim component subjected to thermal loads: boundary conditions and resulting temperature distribution.

# References


[1] Samarskii AA, Vabishchevich PN. Computational heat transfer. The finite difference methodology. Edison, NJ: John Wiley & Sons; 1996.

[2] Reddy JN, Gartling DK. The finite element method in heat transfer and fluid dynamics. 3rd ed. CRC Press; 2010.

[3] Holman JP. Heat transfer. Heat Transf 2010. Available from: http://doi.org/10.1115/1.3246887.

[4] Murthy JY, Mathur SR. Computational heat transfer in complex systems: a review of needs and opportunities. J Heat Transf 2012. Available from: https://doi.org/10.1115/1.4005153.

[5] Shih T-M, Thamire C, Sung C-H, Ren A-L. Literature survey of numerical heat transfer (2000–2009): Part I. Numer Heat Transf Part A Appl 2010. Available from: https://doi.org/10.1080/10407781003656827.

[6] Patankar S. Numerical heat transfer and fluid flow. Ser. Comput. methods Mech. Therm. Sci. Boca Raton, FL: CRC Press; 1980.

[7] Cengel YA. Mass transfer: from fundamentals to modern industrial applications. Heat Mass Transf 2006. Available from: https://doi.org/10.1002/3527609180.

[8] Nakayama A. PC-aided numerical heat transfer and convective flow. Taylor & Francis; 1995.

[9] Hahn DW, Ã--zisik MN. Heat conduction. Edison, NJ: John Wiley & Sons; 2012.

[10] Modest MF. Radiative heat transfer. NL: Academic Press, Elsevier Inc; 2013.

[11] Kreith F, Manglik RM. Principles of heat transfer. Stamford, CT: Cengage Learning, Inc; 2016.

[12] O'Connell JP, Haile JM. Thermodynamics: fundamentals for applications; 2005. Available from: http://doi.org/10.1017/CBO9780511840234.

[13] Honig JM. Thermodynamics; 2014. Available from: http://doi.org/10.1016/B978-0-12-416705-6.00010-5.

[14] Legon AC. Second law of Thermodynamics. Nature 1973. Available from: https://doi.org/10.1038/244431a0.

[15] Soares C. Chapter 19 – Basic design theory. In: Gas turbines. 2nd ed. 2015. p. 913–58. Available from: https://doi.org/10.1016/B978-0-12-410461-7.00019-5.

[16] Reithner R, Müller H. CFD studies for boilers. Comput Fluid Solid Mech 2003;2003:988–91. Available from: https://doi.org/10.1016/B978-008044046-0.50241-4.

[17] Vakkilainen EK. Steam generation from biomass: construction and design of large boilers. Butterworth-Heinemann; 2016.

[18] Lalot S. Heat exchangers: types, design, and applications. Heat Exch. Types, Des. Appl. 2011.

[19] Sánchez M, Clifford B, Nixon JD. Modelling and evaluating a solar pyrolysis system. Renew Energy 2018;116:630–8. Available from: https://doi.org/10.1016/j.renene.2017.10.023.

[20] Wheeler VM, Bader R, Kreider PB, Hangi M, Haussener S, Lipiński W. Modelling of solar thermochemical reaction systems. Sol Energy 2017. Available from: https://doi.org/10.1016/j.solener.2017.07.069.

[21] Canepa E, Builtjes PJH. Thoughts on earth system modeling: from global to regional scale. Earth-Sci Rev 2017;171:456–62. Available from: https://doi.org/10.1016/j.earscirev.2017.06.017.

[22] Newman JN, John N. Marine hydrodynamics. Cambridge, MA: MIT Press; 1977.

[23] Rodríguez I, Pérez-Segarra CD, Lehmkuhl O, Oliva A. Modular object-oriented methodology for the resolution of molten salt storage tanks for CSP plants. Appl Energy 2013;109:402–14. Available from: https://doi.org/10.1016/j.apenergy.2012.11.008.

[24] Gandía LM, Arzamendi G, Diéguez PM, Gandía LM, Arzamendi G, Diéguez PM. Chapter 1 – Renewable hydrogen energy: an overview. In: Renew. Hydrog. Technol. 2013; 1–17. Available from: http://doi.org/10.1016/B978-0-444-56352-1.00001-5.

[25] Marchisio DL, Fox RO. Reacting flows and the interaction between turbulence and chemistry. Ref Modul Chem Mol Sci Chem Eng 2016. Available from: https://doi.org/10.1016/B978-0-12-409547-2.11526-4.

[26] Peksen M, Meric D, Al-Masri A, Stolten D. A 3D multiphysics model and its experimental validation for predicting the mixing and combustion characteristics of an afterburner. Int J Hydrogen Energy 2015;40:9462–72. Available from: https://doi.org/10.1016/j.ijhydene.2015.05.103.

[27] Karim MR, Naser J. Numerical modelling of solid biomass combustion: difficulties in initiating the fixed bed combustion. Energy Procedia 2017;110:390–5. Available from: https://doi.org/10.1016/j.egypro.2017.03.158.

[28] Fureby C, Fureby C. A computational study of combustion instabilities due to vortex shedding. Proc Combust Inst 2000;28:783–91. Available from: https://doi.org/10.1016/S0082-0784(00)80281-7.

[29] Mantzaras J. Chapter Three – Catalytic combustion of hydrogen, challenges, and opportunities. In: Adv. Chem. Eng., vol. 45; 2014; 97–157. Available from: http://doi.org/10.1016/B978-0-12-800422-7.00003-0.

[30] Renard P-H, Thévenin D, Rolon JC, Candel S. Dynamics of flame/vortex interactions. Prog Energy Combust Sci 2000;26:225–82. Available from: https://doi.org/10.1016/S0360-1285(00)00002-2.

[31] Peksen M. 3D thermomechanical behaviour of solid oxide fuel cells operating in different environments. Int J Hydrogen Energy 2013;38:13408–18.

[32] Peksen M. 3D CFD/FEM analysis of thermomechanical long-term behaviour in SOFCs: furnace operation with different fuel gases. Int J Hydrogen Energy 2015;40:12362–9.

[33] Imtiaz M, Hayat T, Alsaedi A, Ahmad B. Convective flow of carbon nanotubes between rotating stretchable disks with thermal radiation effects. Int J Heat Mass Transf 2016;101:948–57. Available from: https://doi.org/10.1016/j.ijheatmasstransfer.2016.05.114.

[34] Karel M, Lund DB. Heat transfer in food. Phys. Princ. Food Preserv 2003. Available from: https://doi.org/10.1201/9780203911792.ch3.

[35] Beckermann C, Viskanta R. Natural convection solid/liquid phase change in porous media. Int J Heat Mass Transf 1988;31:35–46. Available from: https://doi.org/10.1016/0017-9310(88)90220-7.

[36] Nield DA, Bejan A. Convection in porous media; 2013. Available from: http://doi.org/10.1007/978-1-4614-5541-7.

[37] Peksen M, Acar M, Malalasekera W. Transient computational fluid dynamics modelling of the melting process in thermal bonding of porous fibrous media. Proc Inst Mech Eng Part E J Process Mech Eng 2013;227. Available from: https://doi.org/10.1177/0954408912462184.

[38] Liu H, Li B, Chen Z, Zhou T, Zhang Q. Solar radiation properties of common membrane roofs used in building structures. Mater Des 2016;105:268–77. Available from: https://doi.org/10.1016/j.matdes.2016.05.068.

[39] Mao A, Luo J, Li Y. Numerical simulation of thermal behaviors of a clothed human body with evaluation of indoor solar radiation. Appl Therm Eng 2017;117:629–43. Available from: https://doi.org/10.1016/j.applthermaleng.2017.02.071.

[40] Moon JH, Lee JW, Jeong CH, Lee SH. Thermal comfort analysis in a passenger compartment considering the solar radiation effect. Int J Therm Sci 2016;107:77–88. Available from: https://doi.org/10.1016/j.ijthermalsci.2016.03.013.

[41] Colomer G, Chiva J, Lehmkuhl O, Oliva A. Advanced multiphysics modeling of solar tower receivers using object-oriented software and high performance computing platforms. Energy Procedia 2015;69:1231–40. Available from: https://doi.org/10.1016/j.egypro.2015.03.165.

[42] Incropera FP, DeWitt DP, Bergman TL, Lavine AS. Fundamentals of heat and mass transfer, vol. 6. John Wiley & Sons; 2007. Available from: https://doi.org/10.1016/j.applthermaleng.2011.03.022.

[43] Stamnes K, Stamnes JJ. Radiative transfer in coupled environmental systems: an introduction to forward and inverse modeling. NY, USA: John Wiley & Sons; 2016.

[44] Yang K, Sun P, Wang L, Xu J, Zhang L. Modeling and simulations for fluid and rotating structure interactions. Comput Methods Appl Mech Eng 2016;311:788–814. Available from: https://doi.org/10.1016/j.cma.2016.09.020.

[45] Vanyo JP. Rotating fluids in engineering and science. Stoneham, MA, USA: Butterworth-Heinemann; 1993.

[46] Naik SN, Vengadesan S, Prakash KA. Numerical study of fluid flow past a rotating elliptic cylinder. J Fluids Struct 2017;68:15–31. Available from: https://doi.org/10.1016/j.jfluidstructs.2016.09.011.

[47] Peksen M, Acar M, Malalasekera W. Optimisation of machine components in thermal fusion bonding process of porous fibrous media: material optimisation for improved product capacity and energy efficiency. Proc Inst Mech Eng Part E J Process Mech Eng 2014. Available from: https://doi.org/10.1177/0954408914545195.

[48] Ting DS-K., Ting DS-K. Chapter 5 – Turbulence simulations and modeling. In: Basics Eng. Turbul. 2016; 99–118. Available from: http://10.1016/B978-0-12-803970-0.00005-2.

[49] Blazek J, Blazek J. Chapter 7 – Turbulence modeling. In: Comput. Fluid Dyn. Princ. Appl. 2015; 213–52. Available from: http://doi.org/10.1016/B978-0-08-099995-1.00007-5.

[50] Chattopadhyay K, Guthrie RIL. Chapter 4.2 – Turbulence modeling and implementation. In: Treatise Process Metall. 2014; 445–51. Available from: http://doi.org/10.1016/B978-0-08-096984-8.00008-2.

[51] Launder BE, Spalding DB. The numerical computation of turbulent flows. Comput Methods Appl Mech Eng 1974;3:269–89. Available from: https://doi.org/10.1016/0045-7825(74)90029-2.

[52] Dullien FAL. Porous media: fluid transport and pore structure; 1992. Available from: http://doi.org/10.1017/CBO9781107415324.004.

[53] Peksen M, Acar M, Malalasekera W. Computational modelling and experimental validation of the thermal fusion bonding process in porous fibrous media. Proc Inst Mech Eng Part E-J Process Mech Eng 2011;225:173–82.

[54] Bernabé Y, Maineult A. 11.02 – Physics of porous media: fluid flow through porous media. In: Treatise Geophys. 2015; 19–41. Available from: http://doi.org/10.1016/B978-0-444-53802-4.00188-3.

[55] Peksen M, Acar M, Malalasekera W. Computational optimisation of the thermal fusion bonding process in porous fibrous media for improved product capacity and energy efficiency. Proc Inst Mech Eng Part E J Process Mech Eng 2012;226. Available from: https://doi.org/10.1177/0954408912457611.

[56] Whitaker S. Flow in porous media I: a theoretical derivation of Darcy's law. Transp Porous Media 1986;1:3–25. Available from: https://doi.org/10.1007/BF01036523.

[57] Whitaker S. The Forchheimer equation: a theoretical development. Transp Porous Media 1996. Available from: https://doi.org/10.1007/BF00141261.

## 3.3 Problems

**3.1** A sample fin component has been used to cool a chip at steady state and transient conditions. The problem has been used that readers recall the typical application found in most heat transfer course books.

Emphasise has been given to the heat generation that appears analogue in various industrial processes. The fin is made of an aluminium alloy, while the chip is assumed to be made of a special resin. The fin is placed in ambient air at 25°C and is subjected to a heat transfer coefficient of 5 W/m$^2$-K. The schematic of the problem has been illustrated in Fig. 3.P1. Adiabatic conditions can be set for the bottom part, stating an insulated situation. A power of 20 W is generated inside the chip due to electric activity. Thermal conductivities of 0.28 and

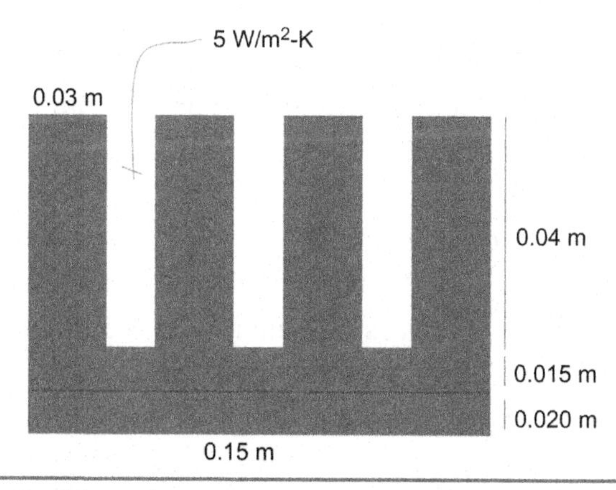


**Figure 3.P1** Problem description.

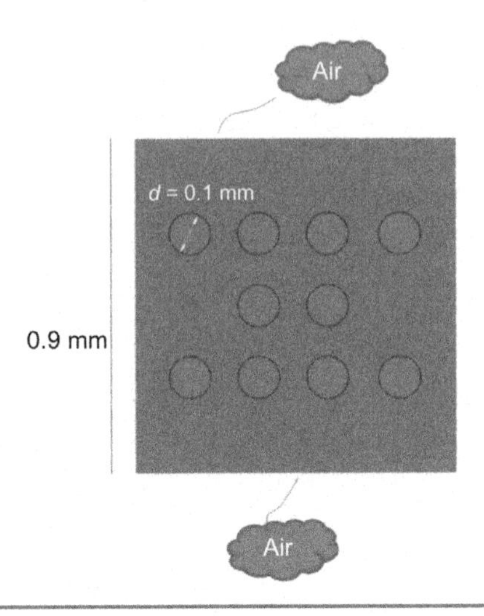


**Figure 3.P2** Problem description.

244 W/m-K are assumed for the resin and the aluminium, respectively.

**a.** Determine the temperature distribution at steady state.

**b.** Calculate the temperature distribution at $t = 100$ s with a time step of 25 s, if the initial temperature has been 75°C. The density of aluminium is 2700 kg/m$^3$ and the specific heat capacity is 4200 J/kg-K. For the resin, density and specific heat capacity are considered as 900 kg/m$^3$ and 1600 J/kg-K, respectively.

Note: The analyst needs to bear in mind that the heat generation is per unit volume, thus the generated amount of energy must be divided by the area of the chip, as it is a two-dimensional case.

**3.2** An arrangement of circular components is attached to a square metal plate of width 0.9 mm on a side. The maximum allowable temperature is 85°C that is assumed to be the initial temperature of the whole assembly. Cooling is provided by air at $T = 25$°C and a heat transfer coefficient of $h = 250$ W/m$^2$-K that is flowing along the top and bottom sides of the plate. The remaining two sides are insulated (Fig. 3.P2).

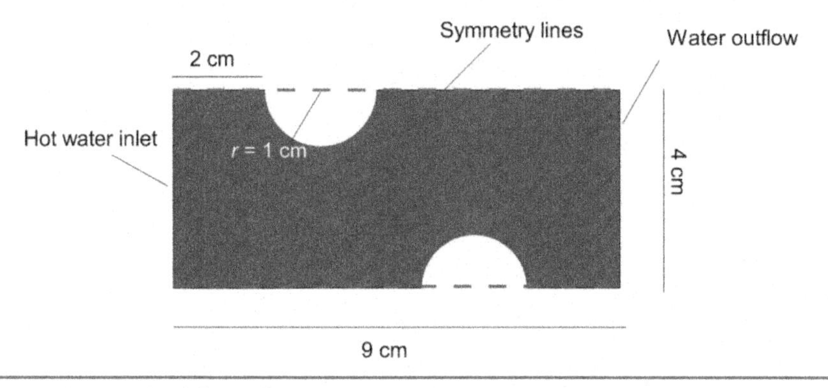


**Figure 3.P3** Problem description.

**a.** Determine the temperature distribution of the components for a cooling time of 5 s. Assuming the properties of the circular components to be $k_{th} = 0.40$ W/m-K, $\rho = 1100$ kg/m$^3$ and $C_p = 1700$ J/kg-K, estimate the time required for the cooling process to reach room temperature.

**b.** Assess the effect of changes in the temperature distribution, if the convection coefficient would be reduced to 100 W/m$^2$-K.

**3.3** Hot water is used to simulate the flow through a two-dimensional channel, containing stationary discs and a flow domain filled with water. The discs are made of high temperature plastic with a $k_d = 0.3$ W/m-K, $\rho_d = 1000$ kg/m$^3$ and $C_{pd} = 1500$ J/kg-K. The fluid flows from the left inlet area into the channel and leaves the domain from the right. The model depicted in Fig. 3.P3 assumes symmetry boundary conditions, due to the periodicity of the problem. The discs are uniformly spaced with a diameter of 2 cm. To capture thermal effects, use a numerical grid containing five layers around each disc. A water inlet velocity of 1 m/s flowing at a temperature of 85°C and outlet atmospheric pressure at the outlet is known.

**a.** Calculate the thermal distribution inside the channel and assess the temperature distribution at the vicinity of the circular discs for a steady state.

**b.** Calculate the transient temperature distribution of the channel, if the fluid inside the channel has an initial temperature of 80°C.

Chapter 4

# Multiphysics Modelling of Structural Components and Materials

Chapter Outline



MULTIPHYSICS modelling of structural components and materials is based on a collection of several physical laws, numerical methods and algorithms to predict the static and dynamic performance of materials and components that are subjected to loading of pure mechanical or coupled nature, including thermal or electrical loads. It has been employed in various fields ranging from earthquake



Multiphysics Modelling. DOI: https://doi.org/10.1016/B978-0-12-811824-5.00004-3

science up to the design and development of fuel cells, wind turbines, combustors, heat exchangers, biomechanics, electronic circuits, etc.

Thereby, answers to the forces required to cause a shape change, the stiffness of the components or critical internal force levels are sought after. In addition, modelling the failure mechanisms are of interest. However, these models are usually experimentally calibrated and do not always perfectly characterise the failure mechanisms. Predicting the fatigue life of a component, the allowable length of crack a component can consider prior fracture or fatigue as well as predicting the creep life are usually of concern. The applications and research within these fields expanded increasingly. To handle all these kind of advanced solid mechanics problems requires a clear and thorough understanding of the theory and application of the important fundamental topics of statics, dynamics and mechanics of materials.

These are required to understand and explain the physical behaviour of the material and components under load and then modelling this behaviour. However, due to the geometrical and mathematical complexity, analytical solutions are most of the time not sufficient to handle these complexities and are valid for relatively simple geometries. Well-known solutions have been used and explained in various textbooks including typical structural components such as joints, beams, trusses or frames, supporting loads and machines containing some moving parts [1–11].

However, the solution will be far more complex when loads and physics of different nature are coupled. Therefore, for complex geometries encompassing additional physics, a numerical solution method such as the finite element method (FEM) is employed [12–21]. Hence, the hallmark of this chapter is that emphasise is placed on the numerically solved solid mechanics, since it will form a suitable basis for the design and analysis of various types of multiphysics engineering systems. The reader is getting used to understand the behaviour of components and materials when subjected to the action of mechanical loads, before combining these effects with loads of different nature.

Many of the engineering applications are concerned with the equilibrium of bodies that are in rest or in motion with a constant velocity—*statics*. On the other hand, in cases where the accelerated motion of bodies is of interest, the subject *dynamics* is considered. Regardless of the equilibrium state of the problem, all analyses require a certain degree of knowledge about the shape of the solid to be analysed. This degree depends on the detail of information that is of interest. Modelling fatigue life or fracture of materials requires great detail as geometrical entities may result in stress concentrations, whereas creep strain or vibration analyses neglect often dimensions under 10% of the macroscopic cross section. In practice this means, if the used model responses well to your questions, then it is sufficient. Otherwise it will serve as a fundamental to build on.

When dealing with the loads, the analyst should carefully understand the type of loads the problem is subjected to and take help from standards such as building codes or requirements for vehicles, processes, etc. These also apply to other fields of multiphysics. Ultimately, the solid mechanics problems, which is the core subject of this chapter are formulated and solved on the basis of the Newton's well known three laws of motion.

Before introducing the procedures to solve the complex behaviour of structural problems, it is useful to remember and comprehend some important aspects that has great influence on the structural simulation predictions.

## 4.1 Material Properties and the Material Behaviour

Any high-tech materials or 'functional' materials (characterised as those materials, which possess particular native properties and functions of their own) have certain properties, as well as show a particular behaviour when subjected to loading of different nature. This is valid for any class of materials such as ceramics, metals or polymers. To analyse their structural response whilst subjected to loading, it is paramount to start to gather information about typical mechanical properties such as elastic module, Poisson's ratio, strength, or gain knowledge about their structural elemental constitution such as being porous, fibrous, ductility, etc. and also determine the force-deformation characteristics of materials under certain loading type and conditions.

From the fundamental courses, it is well known that when describing a material's mechanical behaviour, it is common to plot and learn its behaviour from the experimentally determined stress as a function of the strain curve. Deciding to choose the right equations is the most crucial part of setting up the structural analysis. From the data of a tension or compression test, it is possible to compute various values of stress and strain.

The so-called 'Stress–Strain Curve or Diagram' is important as it provides information about the materials behaviour under load, regardless its physical shape, size or geometry. Each material has a unique line/curve, but for most materials the initial curve is a straight line, reflecting the linear relationship between the stress and strain. This is called the 'Elastic' range. The well-known Hooke's Law only applies in this range.

While most materials have an elastic range, some do not, as we know from basic materials courses [22,23]. As noticed, some basics will be elucidated within this section on purpose, to improve the understanding in approaching and solving solid mechanics problems. To calculate and predict the structural behaviour, it should be bore in mind that several definitions should be understood, in order to ensure the results are the expected ones and how, what to evaluate accurately. Moreover, it must be interpreted what physical meaning those definitions have and how these are mathematically formulated prior use.

Thus it is not surprising that the reader faces again the classical stress–strain diagram of steel. If the material is stressed beyond its elastic limit, it undergoes a permanent deformation. This means the material has yielded and will stretch easily with little additional loading. At the next stage, the material cross section will become slimmer, which is termed as 'necking'. Since the cross-sectional area is changing, the effective or true stress will level up. However, usually the stress calculations will consider the original cross-sectional area (referred as nominal area) and this makes the stress–strain curve appear to go down. So far, pretty known information. But how does the described behaviour affect the targeted numerical predictions?

If we compare an elastic analysis to an elastoplastic analysis, it will be visible that dealing with analysis of an elastic-perfectly plastic material model, i.e., a material that has lost all ability to return to its original shape after deformation gives different results. Fig. 4.1 illustrates the comparison of a linear stress solution (top) from a steel sample compared to a nonlinear solution.

The nonlinear analysis shows that almost the entire cross section of the specimen is subjected to high strain, and a slight increase in

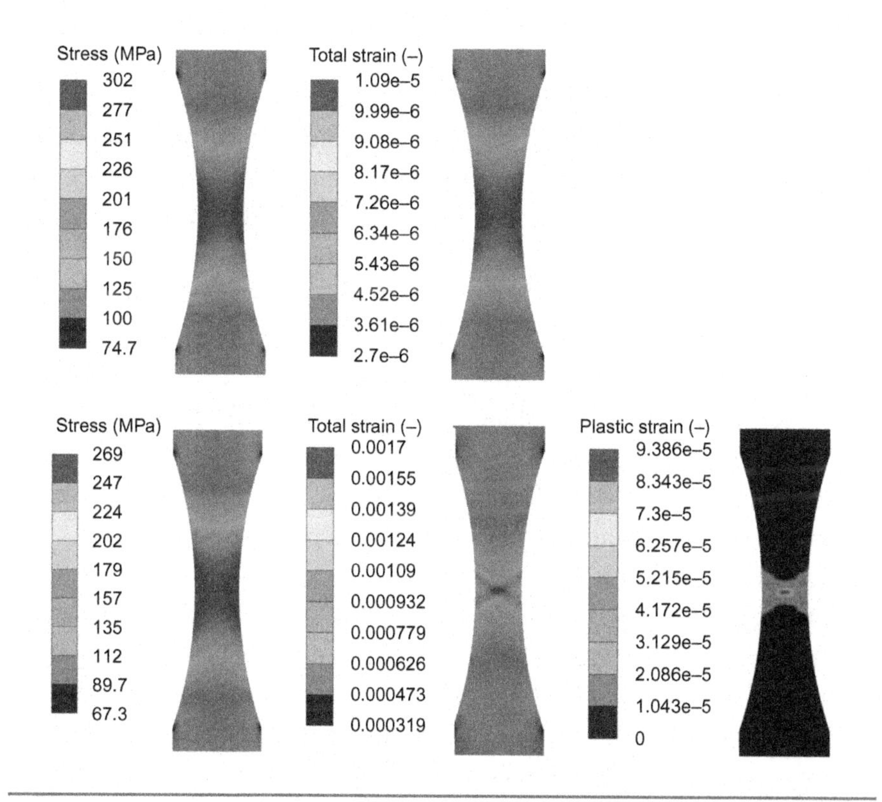


**Figure 4.1** Linear approach (top) versus nonlinear approach (bottom).

the load magnitude would may be cause the cross section to become completely plastic and cause the tested sample to even break. The linear approach shows considerable higher stress values, as well as significant low strain values. Obviously, no plasticity has been considered in the elastic case, thus no contour plot is presented. Ultimately, a difference is present! But where does this difference come from?

In reality one would explain that the material yields and becomes softer, thus the stress increases slower, but in return the strain will be higher. When this effect would not be considered, yielding and the associated conversion of stress into strain would not be present. Let's get into some more detail to shed light on the facts, leading to this difference in the numerical predictions.

The term 'stiffness' causes the fundamental difference between the linear and nonlinear analyses. Stiffness is a property of a material, part or assembly that characterises its response to the applied load. Shape, material type and boundary conditions, i.e. constraints, varying contact and loading situations are all nonlinear effects that change the stiffness of the part (exemplified in Chapter 1: Introduction to Multiphysics Modelling). An I-beam has different stiffness from a channel beam.

An iron profile is less stiff than the same size steel profile. When a structure deforms under a load its stiffness changes attributed to one or more of the factors listed earlier (Fig. 4.2). If it deforms extensively, its shape can change. If the material reaches its failure limit, the material properties will change, as well. On the other hand, if the change in stiffness is small enough, it makes sense to assume that neither the shape nor material properties change at all during the deformation process.

This assumption is the fundamental principle of a *linear analysis*. That means that throughout the entire deformation process, the analysed model retains whatever stiffness it possessed in its undeformed shape, prior to loading. Regardless of how much the model deforms, whether the load gets applied in one step or gradually, the model retains its initial stiffness. Moreover, how high the stress

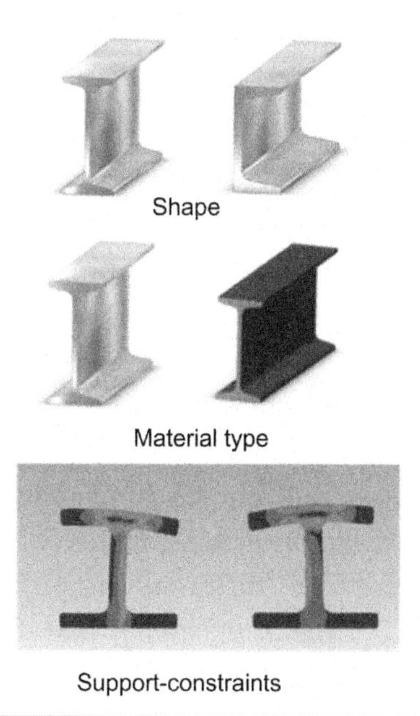


**Figure 4.2** Nonlinear effects influence the stiffness of a part.

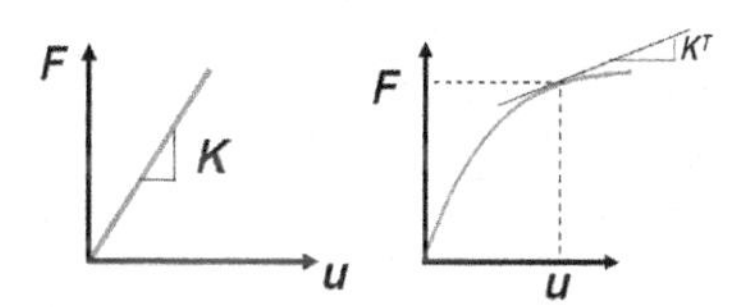


**Figure 4.3** Linear case versus nonlinear case: in the nonlinear case, the stiffness is no longer a constant, *K*; becomes a function of the applied load, $K^T$ (the tangent stiffness). Thus the *F* versus *u* plot for such structures is not a straight line, anymore.

value reaches in the simulation does not matter. Hence, the assumption is that the magnitude, orientation and the distribution of the applied loads remain constant as the structure deforms (Fig. 4.3).

A wide range of problems are considered to be linear because they are loaded in their linear elastic, small deflection range. Thus the nonlinearity effects are too small to detect a difference between a linear and nonlinear solution.

However, the materials behaviour may change due to temperature or a section of the component may lose its stiffness because of buckling or failure where the stiffness changes during the deformation process and must be updated during the computation. Hence, a nonlinear analysis becomes necessary. If the change in stiffness is due to the geometric shape, the nonlinear behaviour is defined as geometric nonlinearity. The complexity and solution time of the problem is affected based on the modelling approach and can be summarised as illustrated in Fig. 4.4.

Based on the utilised material behaviour curve, the numerical solver will calculate a different stiffness, thus the level of the predicted stresses will be different. Therefore reviewing the experimentally determined stress–strain curves is in practice the best way to understand the level of material nonlinearity of the particular problem. In practice, there are different simplified material model choices to be utilised. Linear elasticity that has been applied to metals, ceramics and polymers are widely used.

More advanced models such as hyperelasticity are employed in rubber or foams as they can sustain large dimensional changes. Constitutive equations describing the viscoelastic behaviour comprise usually polymer-based or slow creep containing amorphous solids or materials that show hysteresis during cyclic loading. Rate-independent models used for metals comprise a wide range of sophisticated models, including the prediction of low cycle fatigue or ductile fracture. Viscoplasticity models for instance are suitable for modelling of creep or high-speed machining. There are also very

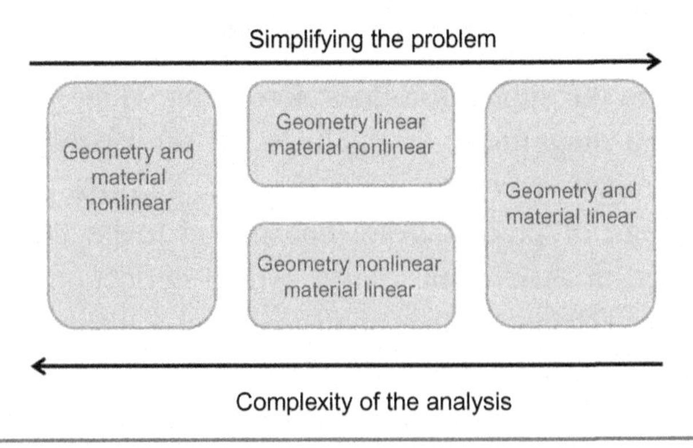


**Figure 4.4** Complexity of a structural component analysis: as the nonlinearity of a problem increases, the more complex it becomes.

specific models predicting, e.g. the plastic flow in single crystals or the crushing in concrete.

It should be noted that the linear analyses give information about the onset of yielding. On the other hand, the nonlinear plasticity analyses are also limited so that they can only predict the onset of fracture, thus the nonlinear material effects such as in plastics can be important to understand the behaviour past the initial yield of the material.

Brittle materials such as cast iron or most of glasses have little inelastic deformation before failure, where a linear approach is generally accepted. However, the majority of materials have some extend of ductility. This will lead to locally material yielding and reduce the stresses compared to what a linear analysis would predict (Fig. 4.5).

Another important effect showing nonlinearities is the so-called follower forces, which mean that the direction of the forces are in motion, adapting to the deformations. As pressure loads are adjusted to retain acting normal to the surface, they are a typical example of forces of this nature. The following example, utilising a simple beam demonstrates the nonlinear effect due to follower forces (Fig. 4.6).

As a part is subjected to loading and deforms, the follower forces will adjust the direction of the loads to ensure they stay normal to the surface. The cantilevered sample beam shown in Fig. 4.6 is loaded with a tip load. The three displacement contour plot results show the response of three different large displacement configurations.

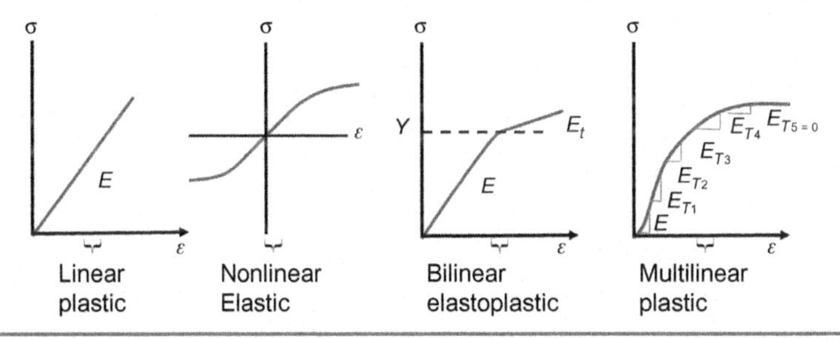


**Figure 4.5** Examples of different material models: as the nonlinearity of a problem increases, the chosen material models have to be adapted according to the material behaviour. Models such as bilinear (with Young's modulus *E* and tangent modulus $E_t$, yield point *Y*), multilinear and nonlinear are widely used.

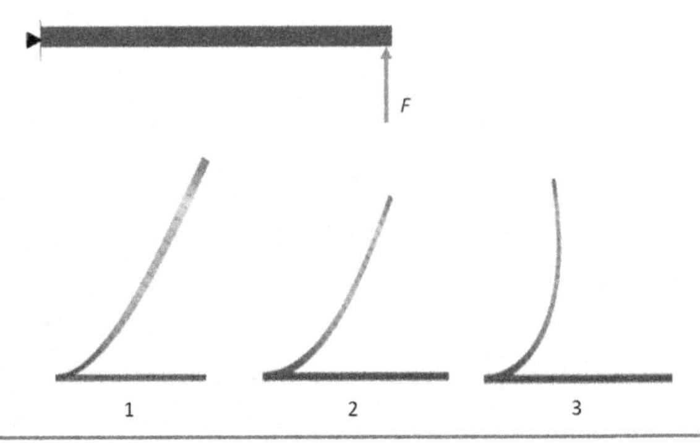


**Figure 4.6** Nonlinear effect of follower forces: displacement contour plot results of three different configurations to predict the deflection of the component.

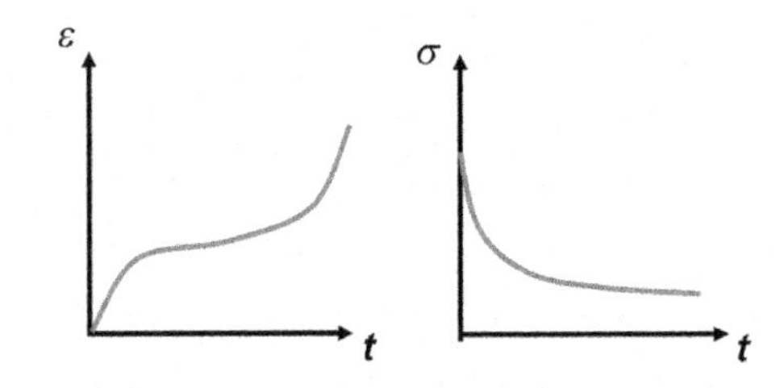


**Figure 4.7** Creep strain versus stress relaxation: under constant applied stress, creep strain increases, whereas stress relaxation occurs due to stress decrease under constant strain.

The first result depicts the configuration where large displacement effects are not considered. The growth is far away from a realistic case. The second analysis shows the results, considering large displacements, but without accounting for the follower forces. The third approach considers both the large displacement effects, as well as the follower forces, which depicts the most accurate case.

In solving multiphysics engineering problems, the long-term behaviour of the materials and components play an important role. A thorough understanding and accurate prediction of this requires to account for the amount of time that a load is applied. This is the case where the creep strain of a certain material needs to be included in the analyses. Viscoplastic constitutive equations are used to predict those cases. These can be interpreted as an extension of the rate-independent plasticity models.

In this section, it has been aimed to give the analyst an understanding about creep strain and its utilisation in numerical analyses, rather than repeating solely the use of elastic or elastoplastic analyses, which are fairly well known.

Creep in general, can be defined as the time-dependent deformation of a material subjected to a certain applied load. The material deforms continuously and permanently at a stress level that is lower than the stress that would cause permanent deformation at room temperature. Creep does not require a higher stress value for additional creep strain to occur. Creep strain is assumed to develop at all nonzero stress values; although it is very often observed in high temperature applications. Moreover, it is also used interchangeably with stress relaxation. However, it should be noted that stress relaxation is a decrease in stress under constant strain, whereas creep is an increase in strain under constant stress (Fig. 4.7).

Creep in materials is usually affected by the change in operating conditions such as of loading and temperature. The creep mechanism is usually different between metals, plastics or low-density porous materials. The creep characteristics of materials affect the strength and the deformation behaviour. Thus it is important to determine and understand the creep behaviour of the used materials and components for their long-term reliability in engineering applications.

Generally speaking, there are two mechanisms of creep based on dislocation-gliding and the diffusional creep [24]. The rate of both is usually limited by diffusion, so both follow an Arrhenius's law. The diffusion becomes appreciable at about 0.3 melting temperature—that is the reason why materials start to creep above this temperature. Therefore employing materials of high melting temperatures are in terms of creep resistance, advantageous.

For the analyst, predicting the creep behaviour is an important task, therefore the underlying creep mechanisms of the used materials such as metals, alloys and deriving the mathematical models has been an important duty [25–27].

Providing a prediction of the life expectancy of materials and structural components requires an initial testing. This is necessary to accurately develop and use mathematical models for the numerical analyses. For this procedure, the dimensional changes of the materials at constant temperature and constant load or stress are accurately measured. In addition, to mimic the thermal conditions, it is common to use an environmental chamber. This is important as to minimise the thermal expansion effects on the sample.

The unloaded specimen is first heated to the desired temperature, following the gage length measurement. The sample load is then applied quickly, but without shocking the material. The emphasis has been to achieve a minimum strain rate under the applied stress and temperature. The deformation over several thousands of hours is of interest compared to classical rupture tests that investigate the time to reach failure in short term. For the analyst, it is important to understand the experimental output of the tests. The rate of the deformation in the measurement is called the creep rate. It is the slope of the line in a Creep Strain versus Time curve.

Creep deformation can undergo three classical stages named as primary, secondary and tertiary. The transient primary period starts at a rapid rate and slows then with time, whereas the secondary stage is the steady region, showing a uniform rate. Within the tertiary stage, the material develops in practice grain boundary voids and shows an accelerated increasing creep rate. This terminates when the material ruptures called as creep rupture.

Sometimes, the material may strain until rupture without exhibiting the secondary stage known as stress rupture. Another term known as creep strength is the minimum nominal stress that produces a given strain rate of the secondary creep subjected to specified thermal loading. Most creep laws focus on the modelling and understanding of the secondary creep.

Based on the experimental measurements, the analyst derives material-specific constant parameters that are used to predict the creep strain behaviour in multiphysics simulations. Fig. 4.8 demonstrates a typical constant temperature, constant stress creep curve.

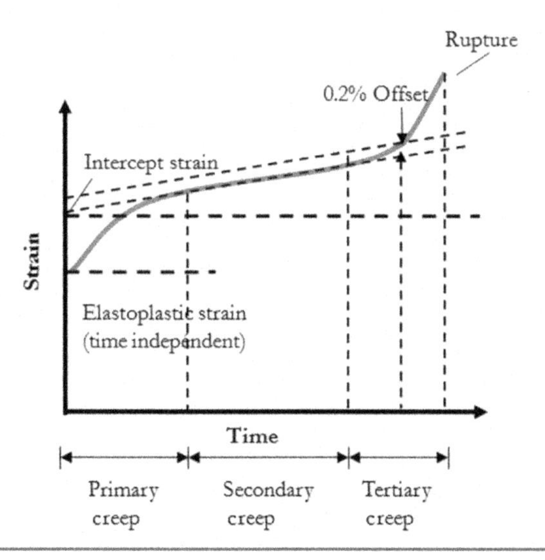


**Figure 4.8** Typical stages of creep strain: primary, secondary and tertiary regions describe the time-independent elastoplastic and time-dependent plastic behaviour.

These require the description of stress states determined in tests such as uniaxial tension, uniaxial compression, torsion and hydrostatic, as the stress fields of most mechanical elements are inhomogeneous. Thus independent creep tests are required to identify the material constants. However, the only exception comprising a homogeneous stress field is the uniaxial stress field in the rod of a constant cross section in tension or compression. This is the reason why a constant cross section is preferred for creep experiments.

There have been a vast amount of proposed creep laws applicable to various engineering materials [28]. One of them is the application of the Larson–Miller law [29], where accelerated tests are used to determine component damage and life by extrapolation methods. The samples are tested at constant temperature and the strain is measured at various stress for a certain time (Fig. 4.9).

The target is to find a design stress by extrapolating the curves to the desired component or material life. However, this only applies when the material does not show microstructure changes that may change the slope of the curves. The following example demonstrates the methodology and how to determine the time necessary for stress

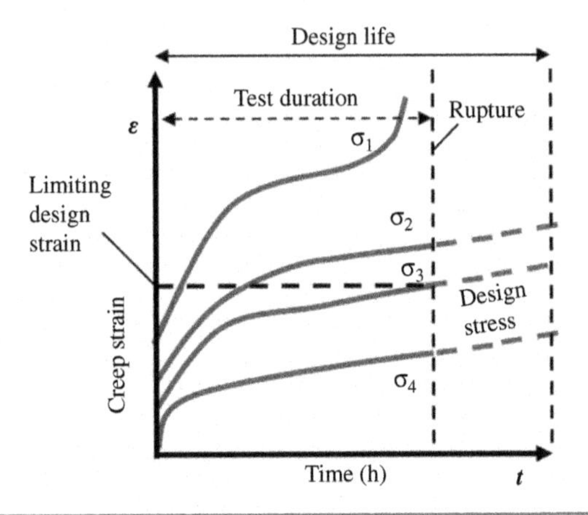


**Figure 4.9** Creep test performed under constant temperature.

rupture to occur at different temperatures. It should be noted that the Larson–Miller parameter $P_{LM}$ utilises the temperature in degrees Rankine (°F + 460) and is expressed as.

$$P_{LM} = (T + 460°)(C_L + \log_t)$$

where $C_L$ is a constant factor usually considered as 20 applicable low alloy steels; alternatively, a factor of 30 is sometimes applied in the case of higher alloy steels in practice. For example if a rod specimen made of low alloy steel is desired to have a life of 10,000 h at say 1000°F and subjected to a load of 100 MPa, what would be the temperature required for the same specimen for an equivalent of 24 h experiment?

From the equation of the $P_{LM}$,

$$\begin{aligned} P_{LM} &= (1000°\text{F} + 460)(20 + \log 10^4) \\ &= 3.504 \times 10^4 \end{aligned}$$

The equivalent temperature can easily found as:

$$\begin{aligned} 3.504 \times 10^4 &= (T + 460)(20 + \log 24) \\ T &= 1179°\text{F} \end{aligned}$$

For the sake of an easy practical understanding, attention has also been drawn to introduce the use of the well-known Norton–Bailey

Odqvist creep model [30–32] that has been widely used in the description of the steady state creep of metals and alloys. The application of the model is demonstrated to the analyst, as the primary and secondary stages of creep play an important role in the design of structural components.

The model can be expressed as:

$$\varepsilon_{cr} = A\sigma^n t^m \tag{4.1}$$

or in rate form as:

$$\dot{\varepsilon}_{cr} = \frac{d\varepsilon}{dt} = A\sigma^n\, mt^{m-1} \tag{4.2}$$

that is called the time hardening formulation of the power law where $A$, $n$ and $m$ are temperature-dependent material constants that are generally independent of stress. While $n$ and $m$ are unitless, the creep strain hardening coefficient, $A$, has units that are consistent with those of the time $t$ and stress $\sigma$. If the main equation is solved for the time $t$ and inserted into the second expression, the strain hardening form is obtained presented as:

$$\dot{\varepsilon}_{cr} = A^{1/m} m\sigma^{n/m} (\varepsilon_{cr})^{(m-1)/m} \tag{4.3}$$

The choice of a creep model depends on the input availability. The parameters need to be extracted from the regression model of the creep data. Since the Norton model is a power law, the equation for the general power law regression fitting is used [30]. Let's demonstrate briefly an example to derive the parameters of the model in its usually known expression as:

$$\dot{\varepsilon}_{cr} = c_0 \sigma^{c1} t^{c2} e^{(-c_3/T)} \tag{4.4}$$

where material-specific constants $C_0$, $C_1$ and $C_2$ are material-dependent creep constants, $C_3$ expresses the material constant defining the creep temperature used as either 0 or 1 based on the temperature dependency. With $C_1 > 1$ and $0 \leq C_2 \leq 1$, $\sigma$ is the applied constant stress at temperature $T$. The letter $t$ refers to the time in hours. In cases where only the secondary creep stage due to having

**TABLE 4.1 Derived Norton Model Parameter for Data Ref. [33]**

| $T$ (°C) | $C_0$ | $C_1$ | $C_3$ | Stress (MPa) | Creep Rate (1/h) |
|---|---|---|---|---|---|
| 450 | $8.352E^{-57}$ | 22.718 | 0.0 | 300 | $1.574E^{-6}$ |
| 500 | $1.376E^{-50}$ | 21.19 | 0.0 | 220 | $5.950E^{-7}$ |
| 550 | $4.566E^{-40}$ | 17.769 | 0.0 | 160 | $6.676E^{-7}$ |
| 600 | $2.490E^{-19}$ | 9.5095 | 0.0 | 100 | $2.601E^{-6}$ |
| 650 | $6.217E^{-12}$ | 6.7473 | 0.0 | 50 | $1.807E^{-6}$ |

an approximately constant material creep rate is present, the hardening term vanishes and the equation is simplified to

$$\dot{\varepsilon}_{cr} = c_0 \sigma^{c1} e^{(-c3/T)} \tag{4.5}$$

The accurate way to extract the constants and use in numerical tools is to apply a bivariate least square fitting procedure and convert the logarithmic data into a linear equation with two variables like $U = ax + by + c$ that are related then to the $C_0$, $C_1$ and $C_3$.

Table 4.1 provides the extracted secondary creep data obtained from the EUROFER 97 steel used in nuclear reactor applications [33]. In this case, attention is drawn again to the use of the Norton model given in Eq. (4.5), where the secondary creep parameters have been of interest. Technically this means that as the secondary creep does not exhibit time- or strain-hardening, the term $t$ and $C_2$ in the equation vanishes (Fig. 4.10).

## 4.2 Geometrical Nonlinearities

The geometric nonlinearity is another aspect that requires attention. It typically becomes an issue when large deformation of a component is valid. These are not meant to be equivalent to large strains such as observed in materials like rubber. Large deformation does not refer to particularly large displacement and rotations, which occur in thin walled structures, either. Because when the stiffness is

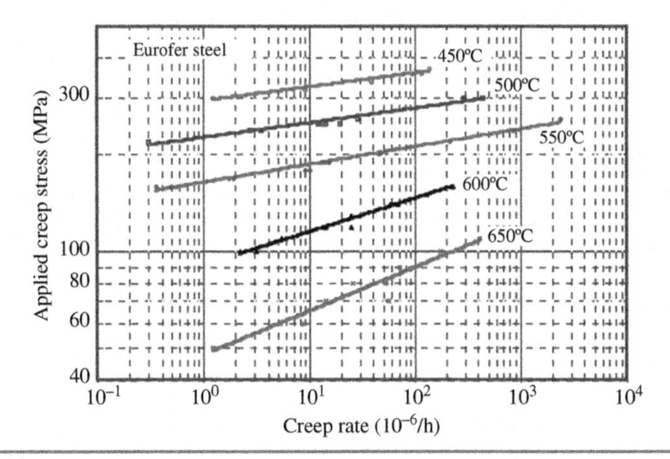


**Figure 4.10** Creep test data obtained from Ref [33].

increased, it is still possible to retain the strains low. But despite this fact, it is still possible to reach high displacements and rotation on the tip of a beam with a satisfactory length. So large strain and large displacements are not equivalent.

Large strains can only occur when large displacement gradients are present, but vice versa is not always the correct case. Because it is possible to achieve large displacement gradients while retaining strains still under 2%. It should be noted that the linear constitutive laws are only valid where large displacement gradients occur but small strains are present.

Large displacement effects can arise due to several different nonlinear properties. Movements or rotations of structural components cause large deflections and should be considered whenever the position of the component becomes significantly different compared to the initial state. The effect of the inclusion of large displacement has been shown in Fig. 4.6 in Section 4.1.

In components that are subjected to rotation, large deflection effects become important. The rigid and flexible deformations need to be combined to predict the accurate geometric stiffness impact on the solution. In practice these become already important by calculation of small angular rotations because the linear approach assumes small displacements where $\sin(\theta) \approx (\theta)$.

Another feature that can be included within this section is the so-called stress stiffening. This is the case when a change in the transverse stiffness of a body is observed when it is subjected to tensile or compression forces along an axial direction. The stiffening is caused by tensile-compression stresses, which result from larger displacements, not by the displacements themselves. They are typically important in components that are short in at least one dimension. In many cases a large deflection solution with stress-stiffening included, converges faster than when it is not accounted for.

Moreover, the prestress effects resulting from transverse changes can change the structure's natural frequencies. Thus considering the stress-stiffening effects computes the stress-induced stiffness matrices that are needed to solve subsequent eigenvalue buckling or pre-stressed modal analyses. Hence, it is important to consider various effects that lead to geometrical nonlinearities that affect the overall quality of the numerical predictions.

Fig. 4.11 exemplifies the stress-stiffening effect, where a prestiffened thin plate is fixed around its perimeter and a uniform pressure load is applied on its surface. This thin walled structure undergoes significant stress stiffening as the component shifts from reacting the load in bending, to reacting the load in-plane. The first analysis shows the deflection where the proper large displacements are considered, whereas the second result depicts the over predicted peak.

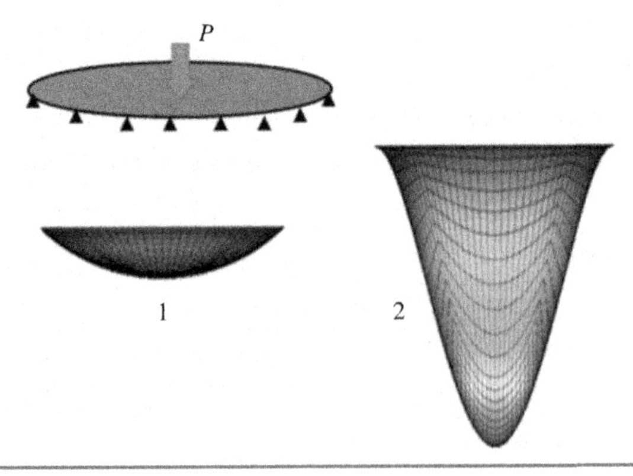


**Figure 4.11** Sample prestiffened thin structure: actual deflection [1]; deflection without considering large deflections [2].

Considering the appropriate material behaviour, material properties, as well as various nonlinearities increases the complexity of the problem as it has been elucidated so far. Dealing with these complexities and solving the structural behaviour of components and materials has been the most common application of the FEM.

In Chapter 1, Introduction to Multiphysics Modelling, several references have been given to the reader to follow up the numerical method and its up to date developments utilised in solving technical problems. The global idea of the FEM used for the structural analysis is to solve a global governing equation in the form of:

$$\underbrace{[K]}_{\substack{\text{Stifness}\\ \text{matrix}}} \cdot \underbrace{\{u\}}_{\text{Displacement}} = \underbrace{\{F\}}_{\text{Force}} \tag{4.6}$$

the displacement is describing the nodal degree of freedom, whereas the force refers to the nodal external force that is applied. The primary unknown to describe the structural behaviour of materials and components are the displacements of the nodes on which the calculations are later performed using the numerical grid. In order to determine the nodal displacements, the global stiffness matrix needs to be formulated. Other quantities, describing the performance of the component under various load types such as strain, stress or forces are then derived from the displacements.

As previously elucidated, the solid body is first to be discretised in several tiny regions called elements. Depending on the problem, these have different shapes such as lines, surfaces and volumes or special types named as shells that are actually two-dimensional surfaces used to model very thin bodies.

In anyway, such as the simplest element represented as lines, the displacement of the end nodes are calculated using iterative or direct solvers. Following this, using constitutive equations such as Hook's Law, the stress is derived using the displacements. In this way, all displacements within each element can be assembled using the direct stiffness method. Thereby, the global stiffness matrix can be related to the global nodal force matrix and the global displacement matrix of the whole structure.

In a nonlinear analysis, the response cannot be predicted directly with a set of linear equations; however, it can be analysed using an iterative series of linear approximations, with corrections such as the Newton–Raphson Method or similar numerical approaches known from basic courses, thus the principle is out of the scope. However, analysts should notice that by increased nonlinearities, convergence of the solution becomes more difficult and methods such as applying loads incrementally or employing convergence enhancement tools of the solvers may be required.

The structural performance of the component is influenced by the loading nature. Therefore different analysis types are considered to determine and characterise the structural behaviour due to various types of loading such as static or dynamic loads. In static loading cases, the component has been subjected to a continuously applied load. They do not cause significant inertia and damping effects. However, the load magnitude can be constant or applied incrementally.

Ultimately what counts is that the response of the structure reaches a steady state. Thus the increments over time represent a counter that identifies the applied steps so that the applied loads and the structure's response are assumed to vary slowly with respect to time. The problem can be both of linear or nonlinear nature, as well as include thermal and fluid-pressure effects.

In the following example, the application of a structural analysis has been demonstrated. A typical disc spring is placed between two rigid bodies as a support. The upper body is subjected to a displacement of 4 mm. An axisymmetrical 2D model has been used (Fig. 4.12). The components are made of ferritic steel with a Young's modulus of 221 GPa and a Poisson's ratio of 0.29.

The materials are assumed to behave linear elastically. Frictionless contact regions are defined on top and bottom of the spring, which are in contact with the top and bottom rigid boundaries. No separation is set for the contact region of the spring and the bottom body.

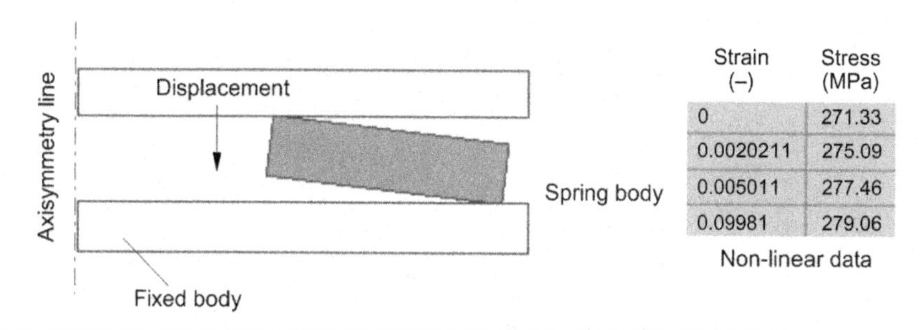


| Strain (–) | Stress (MPa) |
|---|---|
| 0 | 271.33 |
| 0.0020211 | 275.09 |
| 0.005011 | 277.46 |
| 0.09981 | 279.06 |

**Figure 4.12** Disc spring model composed of three components and stress–strain data for the nonlinear analysis.

The results of the linear simulation are compared to the predictions considering a nonlinear multilinear isotropic hardening model with the implemented data tabulated in Fig. 4.12.

Fig. 4.13 depicts the differences in the predicted Von Mises stress distributions for the linear and nonlinear approach.

The deformed and undeformed shapes of the features are illustrated. Note the lower maximum stress value in the nonlinear analysis, which accounts for the material yielding. The stress value is reduced, but in turn the deformation increases and is visible in the plastic strain distribution that occurs at the vicinity where high stress occurs.

When the time scale of the loads becomes important such that the effects of inertia and damping, as well as the applied loads become time dependent, a transient dynamic analysis is required as to determine the time-varying variables of the structural response such as displacements, strains and stresses. In a dynamic analysis, in addition to structural elasticity force, structural inertia and damping are also considered in the equation of motion to equilibrate the dynamic forces. Thus the global form of the Eq. (4.6) is expressed as:

$$\underbrace{\{F(t)\}}_{\text{Force}} = \underbrace{[M]}_{\substack{\text{Mass}\\ \text{matrix}}} \underbrace{\{\ddot{u}\}}_{\substack{\text{Nodal}\\ \text{acceleration}}} + \underbrace{[C]}_{\substack{\text{Damping}\\ \text{matrix}}} \underbrace{\{\dot{u}\}}_{\substack{\text{Nodal}\\ \text{velocity}}} + \underbrace{[K]}_{\substack{\text{Stifness}\\ \text{matrix}}} \underbrace{\{u\}}_{\text{Displacement}} \tag{4.7}$$

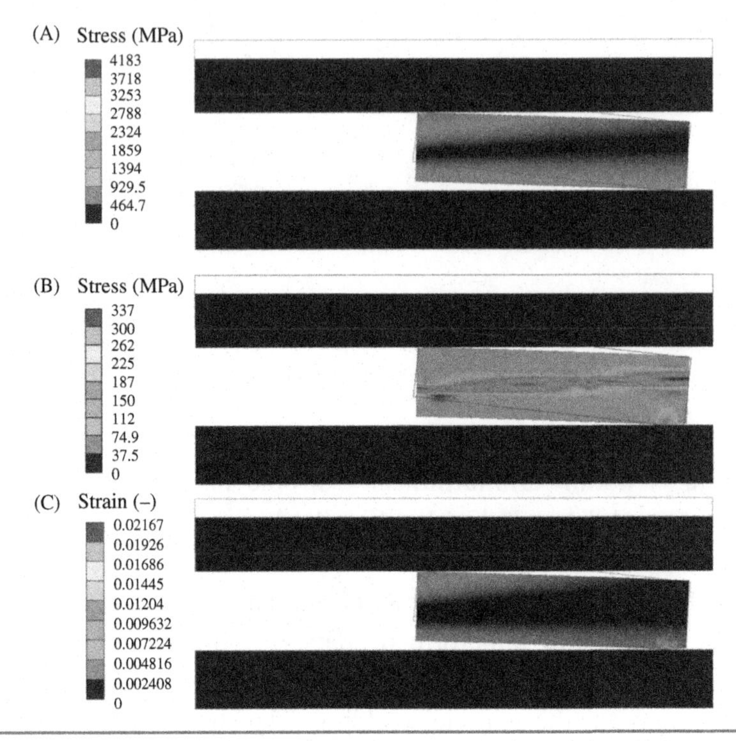


**Figure 4.13** Predicted simulation results: linear approach stress distribution (A); nonlinear approach stress distribution (B) and plastic strain distribution (C).

Buckling and modal analyses are typical applications to determine the structural behaviour of components subjected to dynamic loads. Buckling analyses are used to investigate the instability of components usually subjected to large deflections. The statically applied component load is gradually increased as to determine the level where the utilised component becomes unstable. For example a defined load factor of 4 has been chosen and a static force of 1000 N and a buckling force of 500 has been applied to a structure as compressive force.

The buckling analysis would consider a buckling ultimate load of $1000 + 3 \times 500$ of 2500 N would be applied. When the applied loads are of dynamics nature, then it is important for the component design to understand its vibration characteristics. Using the modal analysis, the natural frequencies and mode shapes are determined.

It should be noted that due to the nature of modal analyses, any nonlinearities in material behaviour are ignored. Therefore it is a good starting point in dynamic analyses.

## 4.3 Rigid Body Analyses

This type of analysis is used to determine the dynamic response of an assembly of rigid bodies linked by joints and springs. You can use this type of analysis to study the kinematics of a robot arm or a crankshaft system, for example. Inputs and outputs are forces, moments, displacements, velocities and accelerations. All parts are rigid such that there are no stresses and strain results produced, only forces, moments, displacements, velocities and accelerations.

The solver is tuned to automatically adjust the time step. Doing it manually is often inefficient and results in longer run times. Viscous damping can be taken into account through springs. Multibody simulation consists of analyzing the dynamic behaviour of a system of interconnected bodies composed of flexible and/or rigid components. The bodies may be constrained with respect to each other via a kinematically admissible set of constraints modelled as joints. These systems can represent an assemblage of rigid and flexible parts of many applications. In any case, the components may undergo large rotation, large displacement and finite strain effects. The subject will not be detailed further.

## 4.4 Stress Concentration, Fracture and Fatigue

As a response of nonlinearities, it is often observed that at the vicinity of cross-section changes, stress distributions receive much higher values than the average value observed over the whole section. This kind of peak stresses particularly present close to holes, grooves, sharp corners and notches are called stress concentrations. From mechanics courses, it is well known that a *stress concentration factor* is defined to consider the ratio of the peak stress to the nominal stress that would arise when the distribution of stress would be uniform expressed as:

$$Kt = \left(\frac{\sigma_{\max}}{\sigma_{\text{nom}}}\right) \tag{4.8}$$

These stresses arise not only in very high values, but also in very low values for the remainder of the section. A broad discussion of stress concentration factors and data can be found detailed in the literature [34–45]. The message for the analyst is to draw attention to the areas where the transmission of the forces occur as their geometrical assumptions or design considerations may influence the results of their predictions that has been discussed in earlier chapters. Sharp transmissions in the direction of the force should be avoided. The transmission from one point to another point must be smoothly.

When notches are necessary in a component, the analyst should bore in mind that removal of material at the vicinity of the notch can significantly alleviate stress concentration effects. Material removal can significantly improve the strength of the component. For practical ease, a typical stress raiser type, i.e. a central single hole in a finite plate is demonstrated as an example. The following plate model is axially loaded. An elastic static analysis has been performed, considering the material as steel. An adaptive grid of 5.0 mm element size has been used. An analytical expression for plane stress conditions has been used to examine the quality of the predicted result. Fig. 4.14 depicts the schematic description of the problem together with the numerically calculated maximum principal stress distribution.

The tensile loading of the plate results in stresses, as the movement is hindered through the applied support on the right hand side of the structure. The resulting stress is not uniform and receives a maximum at the hole vicinity. Note the symmetry in the close up view, indicating the possible use of even a quarter section of the full 3D model. The analyst must bear in mind that in case a quarter section of the geometry is used, the sliced faces must be assigned with

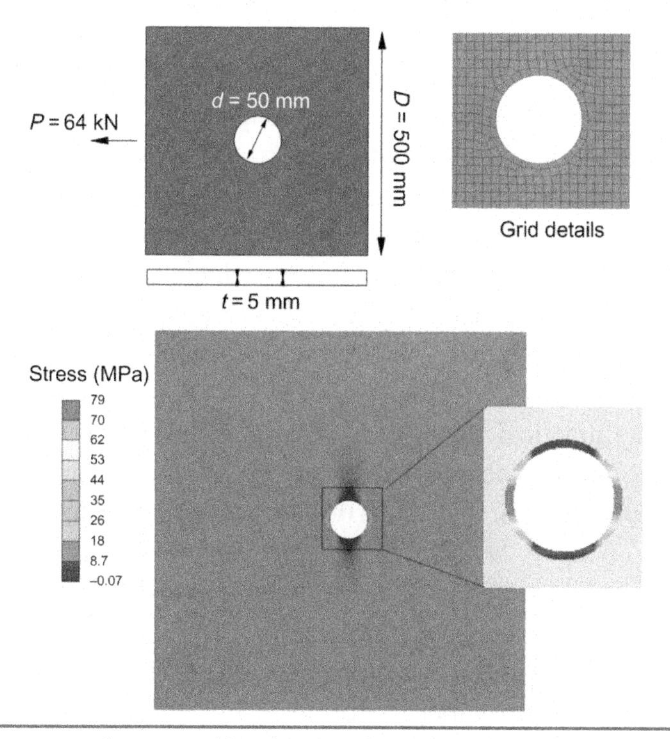


**Figure 4.14** A finite width plate with a central single hole: problem details and the resulting maximum principal stress distribution.

symmetry boundary conditions. The predicted maximum stress is evaluated, using the analytical expressions given as [46]:

$$
\begin{aligned}
&Kt = 3.0 - 3.14\frac{d}{D} + 3.667\left(\frac{d}{D}\right)^2 - 1.527\left(\frac{d}{D}\right)^3 \\
&\sigma_{nom} = \frac{P}{[(D-d)t]} \\
&\sigma_{\max} = K_t\sigma_{nom}
\end{aligned}
\tag{4.9}
$$

Inserting the geometrical values into Eq. (4.9) leads to a stress concentration factor, i.e. $Kt$ of 2.72 and a nominal tension stress of 28.4 MPa. Accordingly, the maximum tension stress at the edge of the hole would be calculated as 77.4 MPa. The numerically predicted maximum stress with a value of 79 MPa predicts a difference of 2.067%, indicating a good agreement.

When optimum structural performance is desired, smaller safety margins are required. To evaluate the strength of cracked structures, the topic known as *fracture* becomes important. Examples, regarding fracture will be demonstrated in Chapter 7, Multiphysics Modelling of High-Performance Materials. When the loading is of cyclic nature, the term *fatigue* is of concern. Indeed, fatigue accounts for 90% of service failures in which the component fails below its yield stress. This depends highly on temperature, type of loading material properties, as well as presence of chemical agents (corrosive fatigue).

Due to its practical importance, the space has been used to demonstrate the fatigue behaviour that may be in more complex coupled processes of concern, particularly, when thermal interactions are considered. Fatigue analyses in components are dealing with the process of cyclic loading evolutions and the associated redistribution of the critical stress region.

The point is that the stress concentration at a particular region is subjected to yielding, which is followed by localised plastic deformation. Ultimately, this proceeds until the resistance of the component decreases and crack initiation occurs. With increasing load cycles the crack increases and finally, permanent damage and failure follows. It is obvious that for the fatigue analysis, a structural analysis is a prerequisite. The analyst is dealing with two distinct types of fatigue i.e., Low-Cycle and High-Cycle. The first one involves cycles (10–100,000) with significant plastic deformation, whereas the latter one is focused on the elastic region and low loads, long life (>100,000 cycles). The low cycles are associated with crack initiation or strain life, while the high cycle fatigue is confined to stress life.

The following example in Fig. 4.15 is used to determine the number of strain cycles that is required for crack initiation. The crack is assumed to be at the point of the maximum stress. A low cycle fatigue analysis has been used to predict the damage sensitivity of the region susceptible to fatigue failure. The used stress field resulted from a high temperature application, thus the components

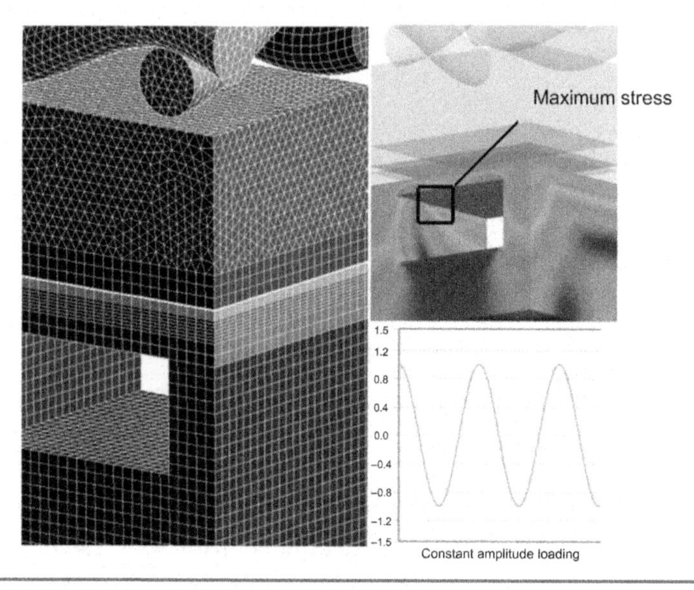


**Figure 4.15** Multiphysics model to predict the fatigue behaviour: numerical grid, maximum stress region and the cyclic load has been depicted.

are expected to show low cycle fatigue behaviour as a predominant failure mode. Since thermal processes are of concern, it is simple to interpret that the fatigue is due to cyclic strain as it is a function of thermal strain. The total strain elastic and plastic is the required to calculate the strain life of the material.

Using this approach, it is possible to determine the damage response of the specimen as a function of the loading history. The damage occurs when the stress–strain field at the critical point changes over time. Therefore it is indispensable to define the way the load is repeating after a single cycle occurs, i.e. the type of fatigue loading determines how the specimen responses over time. The sample has been simulated, using a fully reversed load, which means that the load has been applied and repeated with an equal absolute value and an opposite sign. This is denoted as a load ratio, which in this example refers to a load ratio of $-1$. The amplitude of the load remains constant (Fig. 4.15). The loading is proportional, since only one set of predicted results are sufficient, as the principal stress axes do not change over the time.

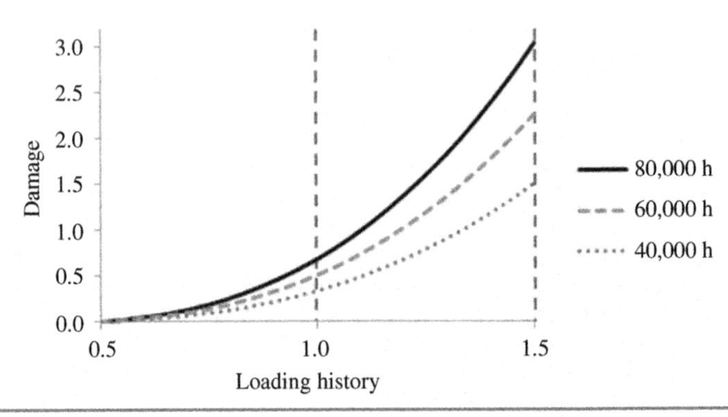


**Figure 4.16** Loading damage analysis for three different design lives.

The damage, which is then to be calculated is defined as the design life divided by the available life. Values over 1 are indicating that specimen damage failure occurs before the design life is reached. Fig. 4.16 presents the results for the sample sensitivity analysis. The assessment has been performed at the critical location given on the critical model point.

The results reveal that for the applied initial strain field load, the specimen retains safe for all cases, because the loading history 1 results for a three time curves under a damage response of 1. However, it becomes visible that when the load is increased, the damage will occur while approaching a loading factor of 1.25 for the 80,000 h case. The loading time until 60,000 h appears to be safe for a loading factor of 1.25; damage will occur beyond this load.

## References

[1] Beer F. Statics and mechanics of materials. McGraw-Hill Education, New York; 2016.

[2] Hibbeler RC. Statics and mechanics of materials Si/engineering mechanics: dynamics Si package. London: Pearson Education, Limited; 2007.

[3] Goodman LE, Warner WH. Statics. CA: Dover Publications, Wadsworth, Belmont; 2012.

[4] Heard WB. Rigid body mechanics: mathematics, physics and applications. New York: John Wiley & Sons; 2008.

[5] Timoshenko S, Young DH. Solutions to problems in statics in engineering mechanics: statics. McGraw-Hill, New York City; 1956.
[6] Goodno BJ, Gere JM. Mechanics of materials, SI edition. CT: Cengage Learning, Stamford; 2017.
[7] Popov EP, Balan TA. Engineering mechanics of solids. NJ: Prentice Hall, Upper Saddle River; 1998.
[8] Young T. Engineering mechanics. FL: The Southeast, Plantation; 1973.
[9] Timoshenko S. Theory of plates & shells 2E. USA: McGraw-Hill Education (India) Pvt Limited; 2015.
[10] Meirovitch L. Elements of vibration analysis. New York: McGraw-Hill; 1975.
[11] Ambrose J. Design of building trusses. New York: John Wiley & Sons; 1994.
[12] Santos TF, Campilho RDSG. Numerical modelling of adhesively-bonded double-lap joints by the eXtended Finite Element Method. Finite Elem Anal Des 2017;133:1–9. Available from: https://doi.org/10.1016/j.finel.2017.05.005.
[13] Zienkiewicz OC, Taylor RL, Fox D. The finite element method for solid and structural mechanics; 2014. Available from: https://doi.org/10.1016/B978-1-85617-634-7.00007-7.
[14] Rust W. Non-linear finite element analysis in structural mechanics; 2015. Available from: https://doi.org/10.1007/978-3-319-13380-5.
[15] Liu GR, Quek SS. The finite element method. MA: Elsevier Science, Butterworth-Heinemann, Waltham; 2014. Available from: https://doi.org/10.1016/B978-0-08-098356-1.00008-4.
[16] Ebbinghaus H-D, Flum J. Finite model theory. DE: Springer Verlag, Berlin; 1999. Available from: https://doi.org/10.1007/3-540-28788-4.
[17] Chen Z. Finite element methods and their applications; 2005. Available from: https://doi.org/10.1007/3-540-28078-2.
[18] Fish J, Belytschko T. A first course in finite elements. New York: John Wiley & Sons; 2007.
[19] Rao SS, Rao SS. Chapter 7 – Numerical solution of finite element equations. In: Finite Elem. Method Eng.; 2011. p. 241–74. Available from: https://doi.org/10.1016/B978-1-85617-661-3.00007-6.
[20] Sadd MH, Sadd MH. Chapter 16 – Numerical finite and boundary element methods. In: Elasticity; 2014. p. 505–29. Available from: https://doi.org/10.1016/B978-0-12-408136-9.00016-7.
[21] Liu GR, Quek SS, Liu GR, Quek SS. Chapter 3 – Fundamentals for finite element method. In: Finite Elem. Method; 2014. p. 43–79. Available from: https://doi.org/10.1016/B978-0-08-098356-1.00003-5.

[22] Roesler J, Harders H, Baeker M. Mechanical behaviour of engineering materials: metals, ceramics, polymers, and composites. Berlin Heidelberg: Springer; 2007.

[23] François D, Pineau A, Zaoui A. Mechanical behaviour of materials. Micro- and macroscopic constitutive behaviour., vol. 1. Netherlands: Springer; 2012.

[24] Ashby MF, Jones DRH, Ashby MF, Jones DRH. Chapter 22 – Mechanisms of creep, and creep-resistant materials. In: Eng. Mater., vol. 1; 2012. p. 337–49. Available from: https://doi.org/10.1016/B978-0-08-096665-6.00022-2.

[25] Ni T, Dong J. Creep behaviors and mechanisms of Inconel718 and Allvac718plus. Mater Sci Eng A 2017;700:406–15. Available from: https://doi.org/10.1016/j.msea.2017.06.032.

[26] Ottosen NS, Ristinmaa M, Ottosen NS, Ristinmaa M. 15 – Creep and viscoplasticity. In: Mech. Const. Model.; 2005. p. 387–421. Available from: https://doi.org/10.1016/B978-008044606-6/50015-1.

[27] He J, Sandström R. Basic modelling of creep rupture in austenitic stainless steels. Theor Appl Fract Mech 2017;89:139–46. Available from: https://doi.org/10.1016/j.tafmec.2017.02.004.

[28] Naumenko K, Altenbach H. Modeling of creep for structural analysis. Berlin Heidelberg: Springer; 2007.

[29] Loghman A, Moradi M. Creep damage and life assessment of thick-walled spherical reactor using Larson–Miller parameter. Int J Press Vessel Pip 2017;151:11–19. Available from: https://doi.org/10.1016/j.ijpvp.2017.02.003.

[30] May DL, Gordon AP, Segletes DS. The application of the Norton-Bailey Law for creep prediction through power law regression; 2013:V07AT26A005.

[31] Peksen M. 3D CFD/FEM analysis of thermomechanical long-term behaviour in SOFCs: furnace operation with different fuel gases. Int J Hydrogen Energy 2015;40:12362–9. Available from: https://doi.org/10.1016/j.ijhydene.2015.07.018.

[32] Peksen M. 3D thermomechanical behaviour of solid oxide fuel cells operating in different environments. Int J Hydrogen Energy 2013;38:13408–18.

[33] Rieth M., Schirra M., Falkenstein A., Graf P., Heger S., Kempe H., et al. Eurofer 97 tensile, charpy, creep and structural tests; 2003. Available from: https://doi.org/0947-8620.

[34] Zeman JL, Rauscher F, Schindler S. Pressure vessel design: the direct route. Kidlington: Elsevier Science; 2006.

[35] Ye XW, Ni YQ, Ko JM. Experimental evaluation of stress concentration factor of welded steel bridge T-joints. J Constr Steel Res 2012;70:78–85. Available from: https://doi.org/10.1016/j.jcsr.2011.10.005.

[36] Zerbst U, Madia M, Vormwald M, Beier HT. Fatigue strength and fracture mechanics – a general perspective. Eng Fract Mech 2017. Available from: https://doi.org/10.1016/j.engfracmech.2017.04.030.

[37] Cao Y, Meng Z, Zhang S, Tian H. FEM study on the stress concentration factors of K-joints with welding residual stress. Appl Ocean Res 2013;43:195–205. Available from: https://doi.org/10.1016/j.apor.2013.09.006.
[38] Chen DH, Masuda K. Estimation of stress concentration due to defects in a honeycomb core. Eng Fract Mech 2017;172:61–72. Available from: https://doi.org/10.1016/j.engfracmech.2017.01.009.
[39] DAAAM International (Vienna). 23rd DAAAM International symposium conference papers. DAAAM International; 2012.
[40] Kiprawi F, Awam FK. Geometrical parameter variations effect on stress concentration factor for tubular KT joints using finite element. Malaysia: Universiti Teknologi; 2014.
[41] Muminovic AJ, Saric I, Repcic N. Analysis of stress concentration factors using different computer software solutions. Procedia Eng 2014;69:609–15. Available from: https://doi.org/10.1016/j.proeng.2014.03.033.
[42] Muminovic AJ, Saric I, Repcic N. Numerical analysis of stress concentration factors. Procedia Eng 2015;100:707–13. Available from: https://doi.org/10.1016/j.proeng.2015.01.423.
[43] Murakami Y, Murakami Y. Chapter 2 – Stress concentration. In: Met. Fatigue 2002; 11–24. Available from: https://doi.org/10.1016/B978-008044064-4/50002-5.
[44] Noda N-A, Shen Y, Takaki R, Akagi D, Ikeda T, Sano Y, et al. Relationship between strain rate concentration factor and stress concentration factor. Theor Appl Fract Mech 2017. Available from: https://doi.org/10.1016/j.tafmec.2017.05.017.
[45] Saini DS, Karmakar D, Ray-Chaudhuri S. A review of stress concentration factors in tubular and non-tubular joints for design of offshore installations. J Ocean Eng Sci 2016;1:186–202. Available from: https://doi.org/10.1016/j.joes.2016.06.006.
[46] Pilkey WD. Formulas for stress, strain, and structural matrices. New York: John Wiley & Sons; 1994.

## 4.5 Problems

**4.1** A plate with a hole in the mid region is subjected to a tensile load of 50 kN (Fig. 4.P1).

**a.** Determine the maximum stress, assuming $E = 200$ GPa, $\nu = 0.28$.

**b.** Calculate the stress concentration factor and the maximum stress in example in 4.1 using an analytical expression.

Compare the two results by determining the difference in percentage.

**c.** The sample in 4.1 shows symmetry in both geometry and loading. By applying the appropriate boundary conditions, solve the same problem using a quadrant of the same model.

**4.2** A square steel plate of 200 mm has a hole of 40 mm at the centre. The plate has been subjected to a mechanical tensile load of 100 kN. Utilising a linear elastic approach, determine the following:

**a.** Von Mises stress distribution of the plate.

**b.** The normal stress components in $x$-, $y$- and $z$-directions.

**c.** Calculate the equivalent strain.

**d.** Calculate the total deformation of the component (Fig. 4.P2).

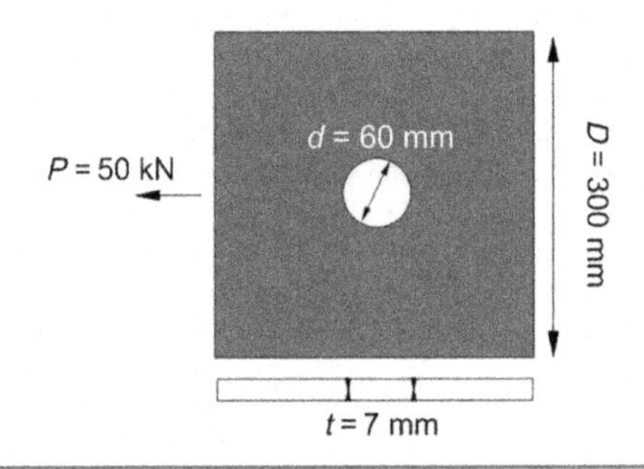


**Figure 4.P1** Problem schematic.

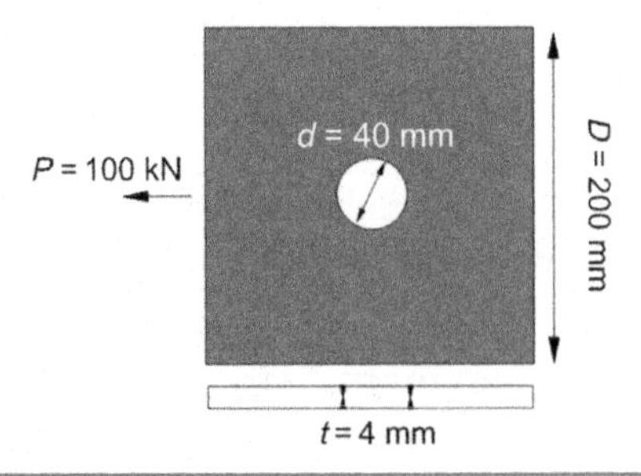


**Figure 4.P2** Problem schematic.

**4.3** Compare your results in Problem 4.2 for each section using an aluminium alloy material with the properties of $E = 72$ GPa, $v = 0.33$.

**4.4** A rod specimen made of low alloy steel is desired to have a life of 15,000 h at 1100°F and is subjected to a load of 150 MPa.

**a.** Calculate the required temperature for the same specimen for an equivalent of 48 h experiment.

**b.** Consider the sample made of high alloy steel with a Larson–Miller parameter *CL* of 30.

**4.5** Consider a plate component according the given dimensions. Assume that the smaller hole is centrally located between the larger hole and the loaded face. The plate shall have elastic properties of $E = 90$ GPa, $v = 0.30$. The applied load is 150 kN (Fig. 4.P3).

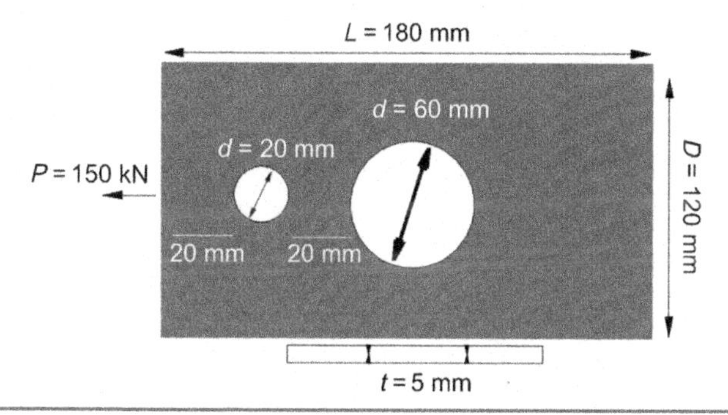


**Figure 4.P3** Problem schematic.

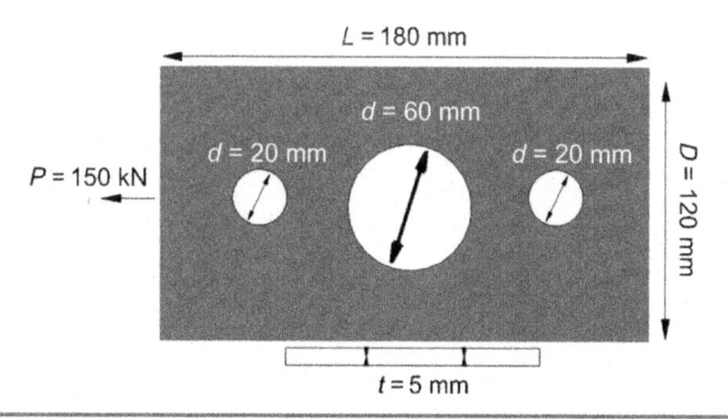


**Figure 4.P4** Problem schematic.

**a.** Determine the stresses around the hole edges and the total deformation.

**b.** By adding one additional hole, observe if the maximum stress achieved in the component could be reduced. Calculate the difference in percentage if the same material properties are used. Consider equal distances between the holes (Fig. 4.P4).

Chapter 5

# Multiphysics Modelling of Interactions in Systems

## Chapter Outline



Throughout this book, it has been elucidated that multiphysics modelling places great emphasis on the interactions between various systems. But first of all, how does a multiphysics analyst define a system? Is it possible to consider generally a system as an integrated set of elements that accomplish a defined objective?

People engaged in different engineering disciplines have different perspectives of what a 'system' is. A software engineer often refers to an integrated set of computer codes as a 'system'. Electrical



Multiphysics Modelling. DOI: https://doi.org/10.1016/B978-0-12-811824-5.00005-5

engineers might refer to complex integrated circuits or an integrated set of electrical units as a 'system'. A photographer may refer a system as his camera system, consisting of interchangeable lenses, the lens focusing mechanism, camera body, flash subsystem, electrical subsystem and power sources (Fig. 5.1).

As can be seen, the term 'system' depends on one's perspective and the 'integrated set of elements that accomplish a defined objective' can be an appropriate definition.

The engineering systems are usually subdivided according to their functions. Apart of major systems like thermal [1,2], physical [3–7], chemical [8–17] or structural components [2,18–21], multiphysics problems are not limited to these and may be concerned even with vital systems such as of environmental [22–25] or biological nature [26–30]. Although each of the multiphysics systems own their unique identities, there is substantial interaction between them that needs attention.

This can be imagined similar to the effects of a volcanic eruption. The geosphere may cause profound effects on the hydrosphere, atmosphere and biosphere. Likewise, in technological systems each constituent may cause direct or indirect effects. Ultimately, two or

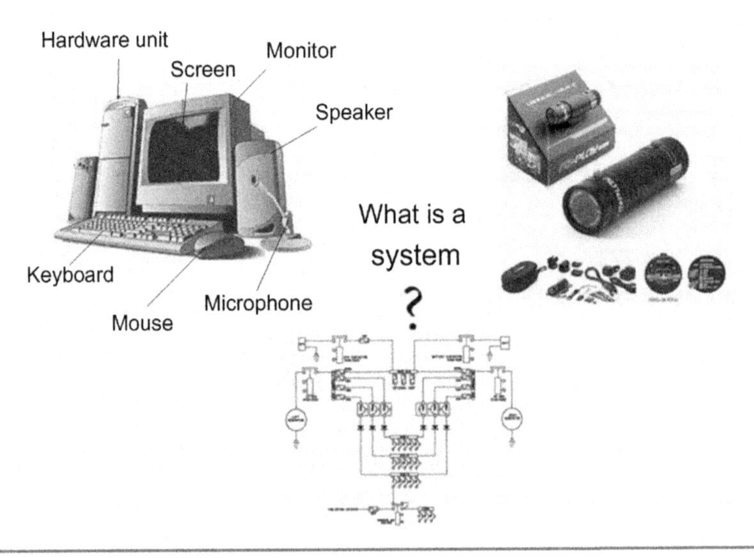


**Figure 5.1** The definition of a system depends on one's perspective.

more events have an effect upon one another. These can be components, materials, processes, etc. that may represent exchanges of matter, energy or information with their surroundings. For a total system development, it is not sufficient to understand and optimise one single component and then construct the remaining parts around it. The contribution and allocation of each component plays a crucial role in the overall design.

Quantifying the interactions between components requires careful modelling strategies. The quantified interactions can capture important propagation patterns of the variables, thus helping to better understand the system and to identify key links and key components that are crucial for the overall system behaviour. The model representation is a very important task. This will determine our degree of details, as well as defines for sure the limits of our computer resources during the calculations.

At the beginning of a concept stage, low-order models may seem to be much more time efficient to first quantify the interactions between components or processes prior to switching to a more detailed multiphysics model. However, in order to develop new concepts into prototypes and ultimately into products, physical system modelling using multiphysics simulation is virtually a necessity.

One of the important features to be taken into account is defining the boundaries of the considered system. This means, the entities that should be included within the modelled system and the ones that are left outside, need to be defined. This should not be confused with a black box considered in the system theory. This would be an approach that contains a process or a collection of processes that transforms inputs into outputs. Multiphysics, however, is interested in details.

Hence, it is important to understand what is inside the black box. This is distinguishing multiphysics modelling from the common complex system modelling approaches. Elements of the system theory can be utilised, however, applied differently. Thus system partitioning or hierarchical modelling can be adopted, but the details within each system partition or elements should be resolved to obtain the details [31].

Interactions within a single component may arise due to the exchange of processes, occurring inside the component. Thus it is important to have knowledge about the size and definition of the system. Whether it is composed of several elements or subcomponents, which is then to be modelled, or a system comprising different processes as well as built up of several components that are interacting. These require more sophisticated multiphysics models such as forming a system architecture, using one single integrated model for the description of multiple views such as concept, analysis and design.

The multiphysics of fluid, thermal and structural components has been the subject in the previous chapters. The current chapter gives a brief demonstration of typical basic interactions observed in fluid–fluid, solid–solid and fluid–solid systems. This will be followed by a large scale system analysis, comprising multiple components.

## 5.1 Fluid–Solid Interaction in Systems

Fluid–solid interaction (FSI) applications involve the coupling of fluid dynamics and structural mechanics disciplines, in general. The basic idea is that a structural component is subjected to hydrodynamic forces exerted by a fluid and ultimately deforms. In turn, the deformed shape of the structure imparts velocity to the fluid domain and changes the flow field. There have been various applications and research within this field [2–4,19–21,32–40].

Based on the level of deformation, a recalculation of the fluid field may or may not be required. In a case where a solid structure is solely moving inside the fluid region and no deformation occurs, this may not be necessary. The following sample demonstrates the case where larger deformations over time are considered and the reader can comprehend the two-way interaction effect. For this purpose, a wire mesh structure is fixed within a fluid domain and fluid is released from an inlet region.

The geometry is shared between the fluid flow and structural solver. Typically, the fluid properties and the flow boundary conditions are specified and for simplicity, the solid material model has been chosen as a standard steel and defined as behaving elastically. Fig. 5.2 illustrates the full FSI problem.

A solution time of 5 s has been demonstrated using a time step of 0.1 s. Another important feature is to identify the interface between the solid and fluid regions. The surfaces of the wire mesh are defined as an interface. Based on the used solver or simulation code the way of calculation may differ. Both the governing equations for fluid flow and mechanics can be assembled and solved simultaneously or a coupling between the fluid and solid is given through passing loads across the fluid–solid interfaces.

Of course the solution can be obtained by calculating each physics separately and then couple them sequentially or simultaneously until an equilibrium is attained. When the solution is over time, each iteration is moving on in time, meanwhile loads and displacements are transferred between the solver. However, the current sample considers the two-way interaction of the FSI system. The fluid produces forces

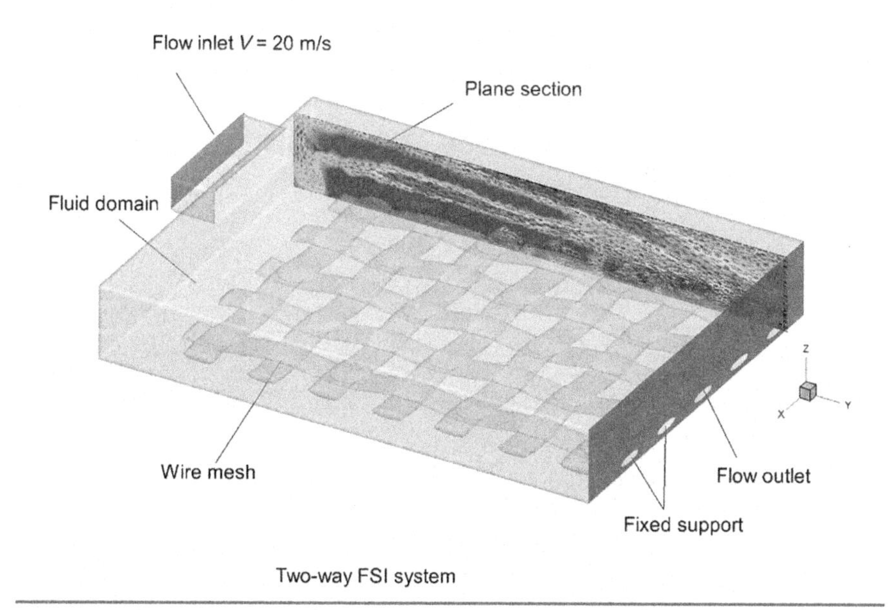


**Figure 5.2** Description of the fully coupled two-way FSI problem. *FSI*, fluid–solid interaction.

on the structure and in turn, the flow field is affected by the displacement of the wire mesh.

The flapping behaviour of the structure induces pressure differences and influences the velocity field over time. Fig. 5.3 demonstrates a sequence of 5 s of the analysis captured on a plane segment. The location of the plane has been given in Fig. 5.2.

As the wire mesh deforms, stresses arise because of the fixed ends. The illustrated Von Mises stress results show the variation of stresses due to the fluctuating flow field. Despite the high stiffness of the metal, still some bending occurs induced by the fluid flow. This movement of the structure affects the flow field over time, as observed by the superimposed velocity vector and contour plots.

Therefore a recalculation of the flow field is required. This data transfer continues throughout the simulation time. If more flexible materials such as rubber or less stiffer polymers would be used for

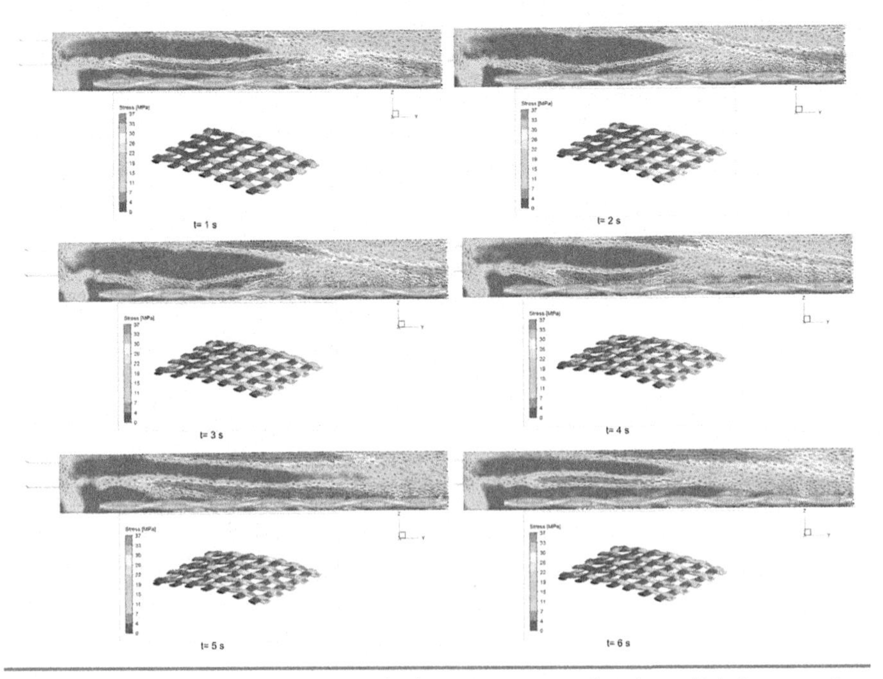


**Figure 5.3** Demonstration of the two-way fluid–solid interaction results: the flow field over time together with its effect on the wire mesh has been predicted.

the analysis, the response of the structure and the recalculated velocity field would be different. However, to demonstrate one of the most challenging multiphysics interactions, the two-way interaction example gives the reader an idea about its application and accurate use.

## 5.2 Solid–Solid Interaction in Systems

Another kind of interaction within multiphysics is the interaction between solids. Solid structures do not always retain motionless or static. In many cases, the contact between multiple domains is different and changes over time. Some are bonded together, slide or even penetrate through each other. In some cases friction is observed, as well. To consider these kind of interactions, advanced modelling techniques are often used.

The solution is obtained such that a maximum number of cycles that refer to time increments is defined. A large number is set to ensure a sufficient simulation time. Usually, a static equilibrium solution is sought after by introducing a damping force proportional to the nodal velocities due to the movement. This targets to damp the lowest mode of oscillation of the static system. In some cases such as in penetration, impact or material fracture, the used elements of the domain degenerate.

A numerical mechanism removes the distorted elements during a simulation. If they are not chosen to be removed then the inertia of the free node is retained and it continues to transfer momentum in subsequent impacts. This happens when its geometric strain exceeds the geometric strain limit. In practice, typical values of 0.5–2 are used.

These kind of analyses demand a uniform mesh size, as the time step of the analysis is calculated based on the smallest element characteristic length. Apart of the type of interaction between the domains, a valid movement of one or more bodies needs to be defined in order to load the model such as in a structural analysis. This may be done by defining a velocity, pressure or gravitational force as boundary condition, for example.

Of course, the system can be prestressed with results from a structural analysis and dynamically solved afterwards. If this is the case, the dynamic nodal position needs to be calculated using the nodal displacement of the static structural analysis. Constraint if any are present will be defined as in normal static cases. When analysing such interactions it is wise to observe and understand the full extent of the damage.

The following example aims to give the reader a quick insight to solid–solid interaction (SSI) example, where an impact-penetration situation has been dynamically solved using multiphysics analysis. A bullet like object illustrated in Fig. 5.4, hits and penetrates through the surface of a composite structure made of steel and nickel parts.

The moving object has a velocity of 650 m/s. The total deformation of the structure has been investigated for a solution time of 0.00019 s. Frictionless contact between the components has been

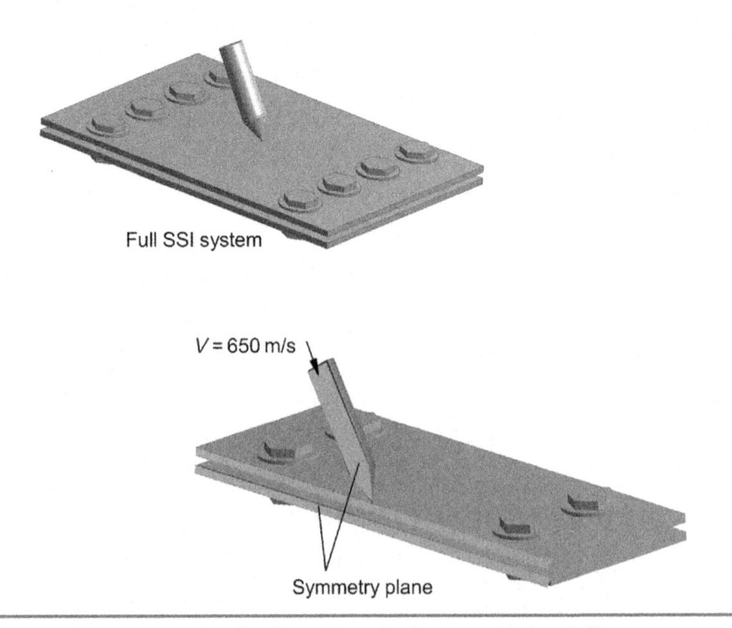


**Figure 5.4** Geometrical details of the model: full geometry and employed symmetrical model with the applied boundary conditions. *SSI*, solid–solid interaction.

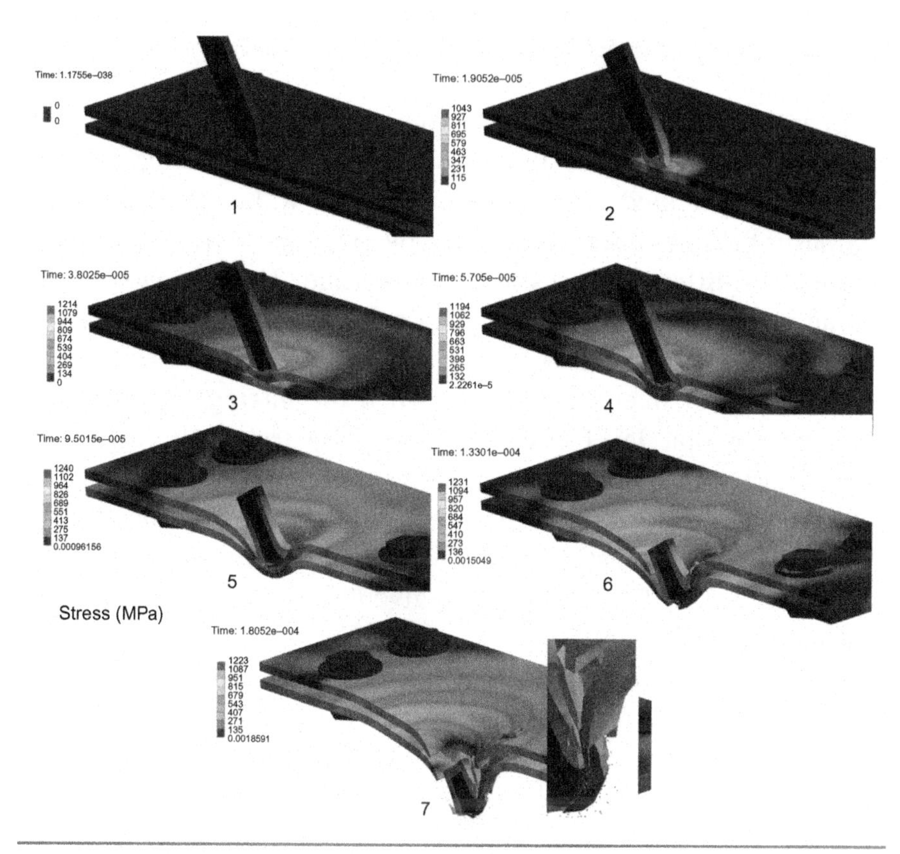


**Figure 5.5** Von Mises stress distribution in the components due to the penetration process.

defined. A strain limit of 0.8 has been used. As this is a typical problem with degradation of elements, erosion has been used retaining inertia of the eroded nodes. Fig. 5.5 demonstrates the stress distribution within the SSI system over time.

The results show the initial impact of the object, deforming the body throughout the sequences over time. For the chosen demonstrative material properties, the Von Mises stresses increase, as the penetration proceeds and resistance occurs. Based on the defined material behaviour the absolute values may of course differ. Mind the eroded sections that is magnified and shown in image 7 of Fig. 5.5.

## 5.3 Fluid–Fluid Interaction in Systems

In Chapter 2, Multiphysics Modelling of Fluid Flow Systems, the multiphysics modelling of fluid flow systems has been elucidated in detail. The last kind of interaction handled within this section is obviously the one observed between two or more fluids of different nature or of different phases of the same fluid. These form mixtures and an interaction occurs. Depending on their mixing level they can be referred as multispecies flow if two or more gases are present. Another type of interaction occurs when the mixing happens on a more macroscopic level where the borders or technically expressed as boundaries are visible. In such cases, these boundaries need to be comprehended by the solver. This is a subject known as multiphase flow that requires attention on its own. However, for the sake of the interactions, a brief touch has been given to the topic that are present in the same domain.

The fluid system consists of a primary phase and all remaining fluids are assumed to be dispersed with a secondary phase. These phases can be categorised within several flow regimes, which is not the focus of this section. Accordingly, discrete gaseous bubbles or droplets in various sizes in a continuous fluid such as in liquid–liquid or liquid–gas mixtures may be present, or immiscible fluids separated by a clearly defined interface such as in free surface flow are examples.

In practice, various applications of different fluid–fluid systems are present; therefore it is important to select an appropriate modelling approach to predict this type of multiphysics interactions.

To demonstrate the modelling of the interactions in a fluid–fluid system (fluid–fluid interaction, FFI) a typical free surface problem has been exemplified. A container has been filled and emptied over time with a liquid substance such as methylene blue into a bucket (Fig. 5.6). The typical liquid–gas interaction between air and the liquid takes place. To solve the multiphase FFI system and to track the motion of the free surface between the two fluids, a volume of fluids [41–45] model has been utilised.

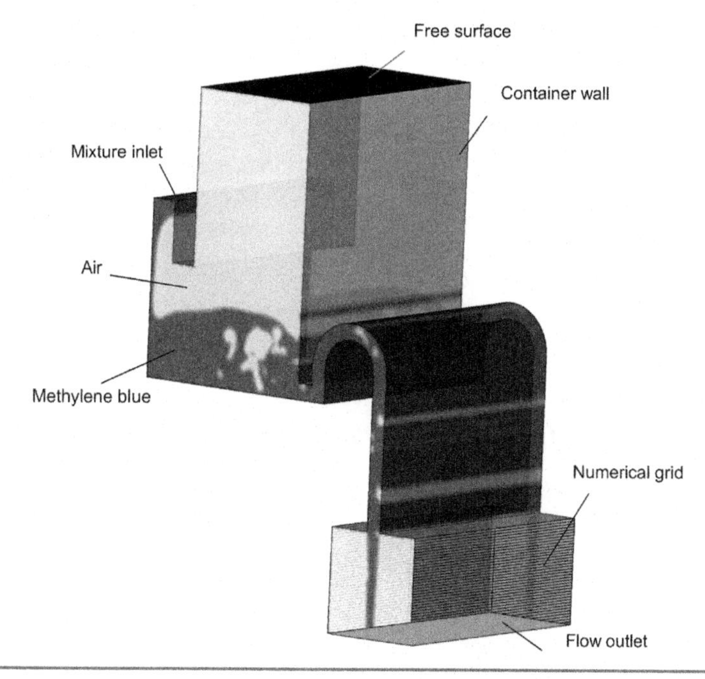


**Figure 5.6** Problem description of the fluid–fluid interaction system.

The liquid level rises until it starts to flow out of the container until it is empty. The gravitational acceleration is enabled. The two phases of air and methylene need to be defined. As the methylene blue is of interest, it is defined as the primary phase, where as air refers to the second one. It is useful to define a surface tension as the interaction between the two fluids. A constant value of 0.03 N/m has been used, for instance.

A turbulence intensity of 6% has been applied to a hydraulic diameter of 2 cm. The $k$-$\varepsilon$ turbulence model has been utilised. The inlet region contains a mass flow rate of 0.5 kg/s methylene blue. The outlet and ambient can be defined as mixture with a backflow volume fraction of 1 for the air. Usually this kind of problems initialise with a volume fraction of 1 for air to consider that the container is first filled with air.

To find an appropriate initial condition the methylene blue needs to be initialised as well. Therefore a region can be patched (a few

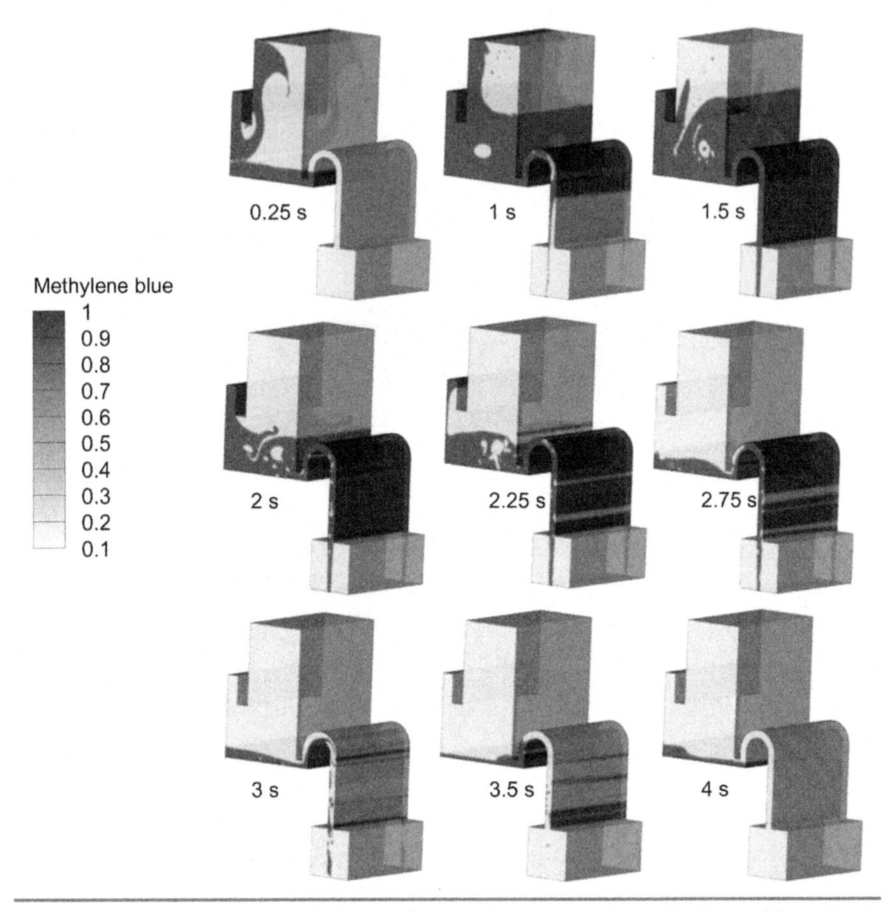


**Figure 5.7** Contours of the volume fraction of methylene blue over time.

cm) as zero volume fraction for air, thus a partial region to be filled with methylene blue that will be sufficient for initialisation of the problem. A solution with a time step of 0.2 s has been performed for 4 s. Fig. 5.7 illustrates the methylene blue volume fraction over time.

## 5.4 3D Multiphysics Modelling of a System Architecture

In this section, a system architecture, comprising of several components and different processes is introduced. An auxiliary

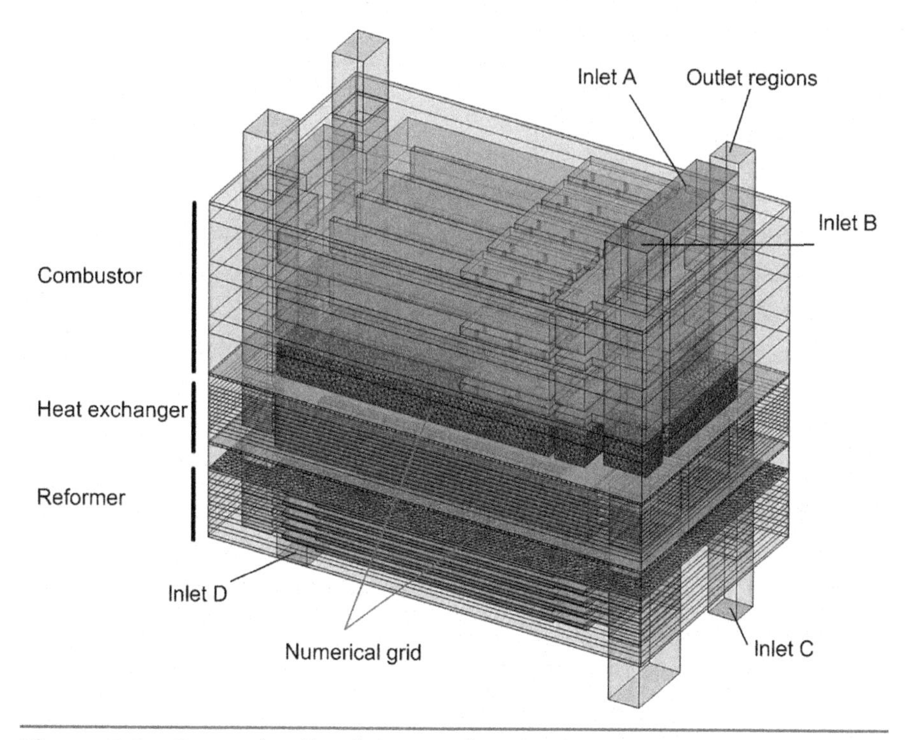


**Figure 5.8** Some details of the used multiphysics model.

system assembly encompassing a reformer, a combustor and a heat exchanger is set. The configuration is shown in Fig. 5.8. The architecture model is a single integrated multiphysics model that solves for the complex turbulent fluid flow, heat transfer, chemically reactive species transport, as well as the combustion.

In this example emphasis has been placed on a detailed understanding about the interrelationship between components and subcomponents. The architecture model provides information about a detailed inventory of fluid flow, heat transfer and chemically reactions that network together. This conveys the multiphysics content of the elements of the system, the relationship among them and the rules governing those relationships that would help for further concept planning and give guidelines governing the design and successful operation.

### 5.4.1 Single Architecture Model and Assumptions

The employed system assembly may serve in practice for the operation of a fuel cell. The combustor is used to recover the heat of incompletely burnt fuel released from the fuel cell. These left-over gases retain enthalpy and are used by burning with excess air. The supplied gases from the fuel cell component are already considered as inlet or outlet boundary conditions that flow into the system. The process gases are kept on operation temperature using a heat exchanger.

The released heat of the combustor is transferred through the walls and the convected gases that are fed into a heat exchanger, or is split in particular ratios, feeding both the reformer and a heat exchanger component. This would then be used for heating the reformer component within the fuel cell system.

The fuel gases are assumed to flow into the system, mimicking the released gases from a fuel cell. The combustor is designed of several layers each made of two parts. In one section, the fuel is introduced through manifolds into a chamber. There, air meets the fuel and combustion takes place. After the reaction, the mixed gas is transported to the reformer and the heat exchanger. The fluid inlet regions for the air and fuel gases are specified as mass flow rates. Fluid temperatures and species concentrations are set, according to the given data in Table 5.1. No-slip conditions are applied at wall

**TABLE 5.1** Used Inlet Boundary Conditions for the Sample

| | Inlet A | Inlet B | Inlet C | Inlet D |
|---|---|---|---|---|
| Temperature (K) | 954 | 973 | 423 | 311 |
| Mass flow rate (kg/s) | 0.01296 | 0.00135 | 0.00079 | 0.01352 |
| Turbulent Intensity (%) | 7.87 | 9.9 | 9.92 | 6.9 |
| Species mass fraction (kg/kg) | Ar 0.0139<br>$O_2$ 0.1989 | $H_2$ 0.0185<br>$CH_4$ 0.0533<br>$H_2O$ 0.5634<br>CO 0.0119<br>$CO_2$ 0.3392 | $CH_4$ 0.31<br>$H_2O$ 0.69 | Ar 0.0132<br>$O_2$ 0.232 |

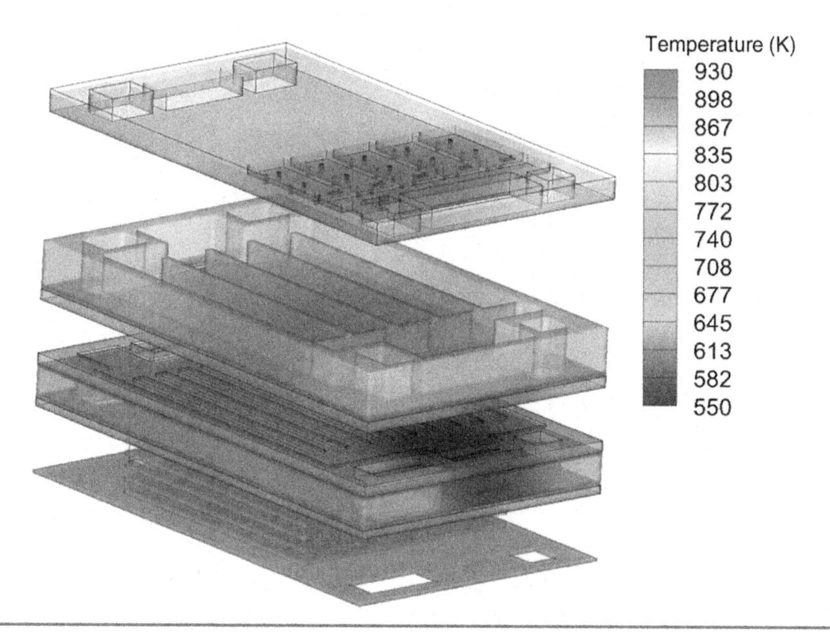


**Figure 5.9** Temperature distribution of the system: the integrated architecture model enables a detailed visualisation of the regions of interest.

boundaries. To derive the turbulence intensity, hydraulic diameters of the inlets are used. Species sum up to 1, thus the remaining concentration departing from 1 is dedicated to $N_2$.

Fig. 5.9 illustrates the detailed predictions of the thermal distribution of the assembly.

The detailed example results illustrate the interacting components and the exchange of energy between the bottom part of the combustor and the cooler reformer component. The hottest zones are as expected inside the chamber region, where the combustion of species takes place. Such an integrated multiphysics model can be used to intensively improve the understanding of specific data.

For instance, if the combustor component is considered, Fig. 5.10 illustrates the thermofluid flow analysis of the region where the jets of the combustion chamber are located. Different postprocessing opportunities enable the analyst to cut slices out of the solution to focus on desired aspects.

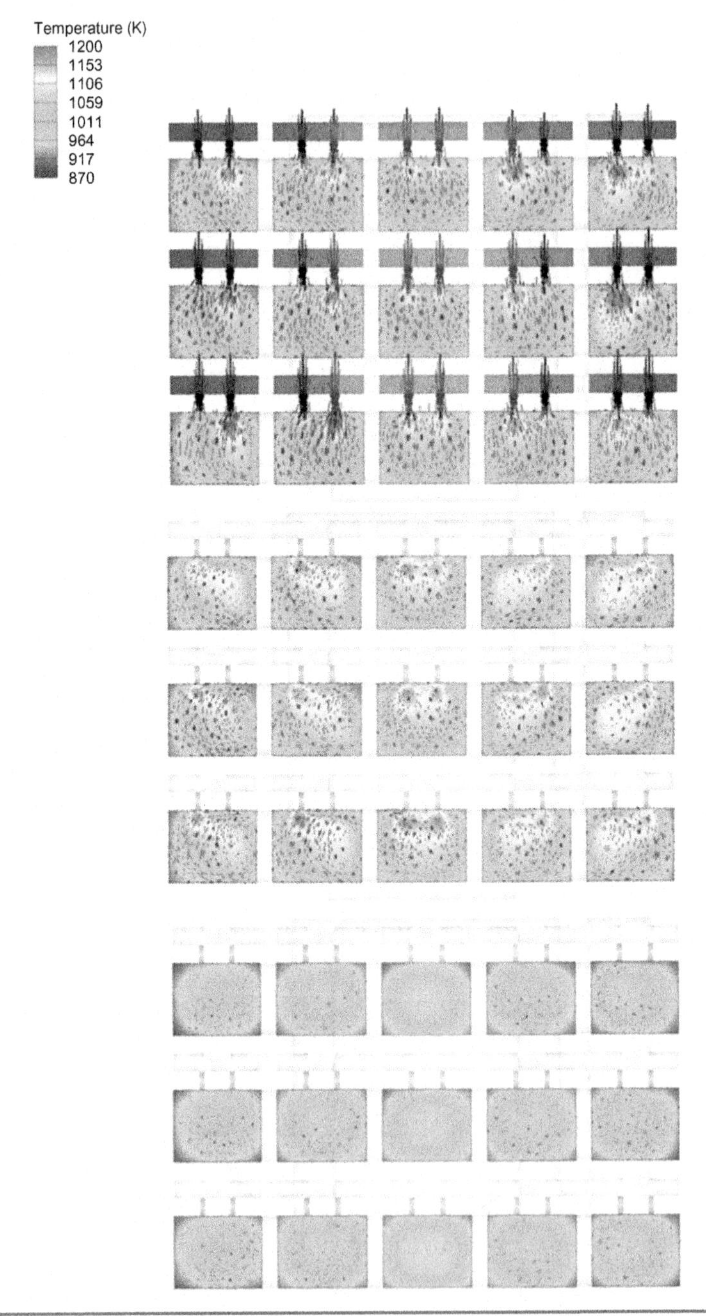


**Figure 5.10** Flow field vectors and thermal distribution at the vicinity of combustion jets.

Accordingly, it can be seen that within the chamber, different slices show that local temperature differences occur. The bottom square section is the region where the fuel and air is mixing. Due to different turbulence-chemical effects, it is visible that predictions about the flame shape, temperature and size can be obtained. Thus it is possible to make a detailed mixing and combustion analysis. Likewise, it is possible to extract extensive information about the species transport to improve the understanding of the transport behaviour, including species concentrations, mixing and reacting nature of the constituents to further improve the understanding (Fig. 5.11).

The methane concentration that occupies the fuel side is reducing as soon as methane fuel flows inside the jets and mixes with the air that is flowing at the bottom part of the chamber. Likewise, the oxygen species mass fraction shown on the right side of the illustration demonstrates the reduction of oxygen mass fraction while approaching the jet region, as it is used to combust the methane that is released into the chamber. The example results demonstrate the power of an integrated multiphysics model to understand the details inside a system and to shed light on the relationship among multiple components and their interactions.

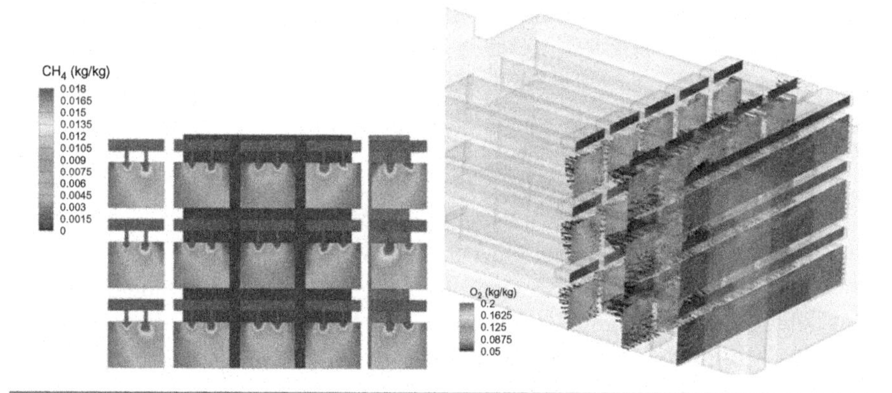


**Figure 5.11** Example of the species concentration distribution inside the combustion chamber: the methane and oxygen mass fractions.

## References

[1] Mantzaras J. Chapter Three – Catalytic combustion of hydrogen, challenges, and opportunities. In: Adv. Chem. Eng., vol. 45; 2014. 97–157. Available from: https://doi.org/10.1016/B978-0-12-800422-7.00003-0.

[2] Bordère S, Caltagirone J-P. A multi-physics and multi-time scale approach for modeling fluid–solid interaction and heat transfer. Comput Struct 2016;164:38–52. Available from: https://doi.org/10.1016/j.compstruc.2015.10.009.

[3] Paidoussis MP, Paidoussis MP. Chapter 2 – Concepts, definitions and methods in fluid-structure interactions. In: Fluid-Structure Interact.; 2014. 7–62. Available from: https://doi.org/10.1016/B978-0-12-397312-2.00002-8.

[4] Parameshwaran R, Dhulipalla SJ, Yendluri DR. Fluid-structure interactions and flow induced vibrations: a review. Procedia Eng 2016;144:1286–93. Available from: https://doi.org/10.1016/j.proeng.2016.05.124.

[5] Bernabé Y, Maineult A. 11.02 – Physics of porous media: fluid flow through porous media. In: Treatise Geophys.; 2015. 19–41. Available from: https://doi.org/10.1016/B978-0-444-53802-4.00188-3.

[6] Chattopadhyay K, Guthrie RIL. Chapter 4.6 – Single phase, two phase, and multiphase flows, and methods to model these flows. In: Treatise Process Metall.; 2014. 527–53. Available from: https://doi.org/10.1016/B978-0-08-096984-8.00012-4.

[7] Cheng L. 3 – Flow patterns and bubble growth in microchannels. In: Microchannel Phase Chang. Transp. Phenom.; 2016. 91–140. Available from: https://doi.org/10.1016/B978-0-12-804318-9.00003-0.

[8] Wheeler VM, Bader R, Kreider PB, Hangi M, Haussener S, Lipiński W. Modelling of solar thermochemical reaction systems. Sol Energy 2017. Available from: https://doi.org/10.1016/j.solener.2017.07.069.

[9] Marchisio DL, Fox RO. Reacting flows and the interaction between turbulence and chemistry. Ref Modul Chem Mol Sci Chem Eng 2016. Available from: https://doi.org/10.1016/B978-0-12-409547-2.11526-4.

[10] Murray P, Carey GF. Viscous flow and transport with moving free and reactive surfaces. Int J Numer Methods Eng 1990;30:1181–94. Available from: https://doi.org/10.1002/nme.1620300607.

[11] Andrianopoulos E, Korre A, Durucan S, Franzsen S. Coupled thermo-mechanical-chemical modelling of underground coal gasification. Comput Aided Chem Eng 2016;38:1069–74. Available from: https://doi.org/10.1016/B978-0-444-63428-3.50183-1.

[12] Piccolo C, Shaw A, Hodges G, Piccione PM, O'Connell JP, Gani R. A framework for the design of reacting systems with phase transfer catalysis. Comput Aided Chem Eng 2012;30:757–61. Available from: https://doi.org/10.1016/B978-0-444-59520-1.50010-5.

[13] Lu P, Binita B, Barton PI, Green WH. Reduced models for adaptive chemistry simulation of reacting flows. Comput Fluid Solid Mech 2003;2003:1422–5. Available from: https://doi.org/10.1016/B978-008044046-0.50348-1.

[14] Lefantzi S, Ray J. A component-based scientific toolkit for reacting flows. Comput Fluid Solid Mech 2003;2003:1401–5. Available from: https://doi.org/10.1016/B978-008044046-0.50343-2.

[15] Foley HC, Foley HC. Chapter 7 – Reacting systems—kinetics and batch reactors. In: Introd. to Chem. Eng. Anal. Using Math. 2002; 297–361. doi:10.1016/B978-012261912-0/50009-X.

[16] Carter JG, Cokljat D, Blake RJ, Westwood MJ. Computation of chemically reacting flow on parallel systems. Parallel Comput Fluid Dyn 1996;1995:113–20. Available from: https://doi.org/10.1016/B978-044482322-9/50068-2.

[17] Boyadjiev CB, Babak VN, Boyadjiev CB, Babak VN. PART 3 – Chemically reacting gas-liquid systems. Non-Linear Mass Transf Hydrodyn Stab 2000;171–223. Available from: https://doi.org/10.1016/B978-044450428-9/50004-1.

[18] Breuer M, De Nayer G, Münsch M, Gallinger T, Wüchner R. Fluid–structure interaction using a partitioned semi-implicit predictor–corrector coupling scheme for the application of large-eddy simulation. J Fluids Struct 2012;29:107–30. Available from: https://doi.org/10.1016/j.jfluidstructs.2011.09.003.

[19] Casoni E, Houzeaux G, Vázquez M. Parallel aspects of fluid-structure interaction. Procedia Eng 2013;61:117–21. Available from: https://doi.org/10.1016/j.proeng.2013.07.103.

[20] Hou G, Wang J, Layton A. Numerical methods for fluid-structure interaction – a review. Commun Comput Phys 2012;12:337–77. Available from: https://doi.org/10.4208/cicp.291210.290411s.

[21] Wick T. Fluid-structure interactions using different mesh motion techniques. Comput Struct 2011;89:1456–67. Available from: https://doi.org/10.1016/j.compstruc.2011.02.019.

[22] Chaduvula U, Patel D, Gopalakrishnan N. Fluid-structure-soil interaction effects on seismic behaviour of elevated water tanks. Procedia Eng 2013;51:84–91. Available from: https://doi.org/10.1016/j.proeng.2013.01.014.

[23] Abdi DS, Bitsuamlak GT. Wind flow simulations on idealized and real complex terrain using various turbulence models. Adv Eng Softw 2014;75:30–41. Available from: https://doi.org/10.1016/j.advengsoft.2014.05.002.

[24] Canepa E, Builtjes PJH. Thoughts on earth system modeling: from global to regional scale. Earth-Sci Rev 2017;171:456–62. Available from: https://doi.org/10.1016/j.earscirev.2017.06.017.

[25] Geyer P. Systems modelling for sustainable building design. Adv Eng Informatics 2012;26:656–68. Available from: https://doi.org/10.1016/j.aei.2012.04.005.
[26] Luraghi G, Wu W, De Gaetano F, Rodriguez Matas JF, Moggridge GD, Serrani M, et al. Evaluation of an aortic valve prosthesis: fluid-structure interaction or structural simulation? J Biomech 2017;58:45–51. Available from: https://doi.org/10.1016/j.jbiomech.2017.04.004.
[27] Lanzas C, Chen S. Complex system modelling for veterinary epidemiology. Prev Vet Med 2015;118:207–14. Available from: https://doi.org/10.1016/j.prevetmed.2014.09.012.
[28] Bardini R, Politano G, Benso A, Di Carlo S. Multi-level and hybrid modelling approaches for systems biology. Comput Struct Biotechnol J 2017;15:396–402. Available from: https://doi.org/10.1016/j.csbj.2017.07.005.
[29] Sun Z, Chaichana T. A systematic review of computational fluid dynamics in type B aortic dissection, vol. 210; 2016. Available from: https://doi.org/10.1016/j.ijcard.2016.02.099.
[30] Taylor CA, Hughes TJR, Zarins CK. Finite element modeling of blood flow in arteries. Comput Methods Appl Mech Eng 1998;158:155–96. Available from: https://doi.org/10.1016/S0045-7825(98)80008-X.
[31] Peksen M, Peters R, Blum L, Stolten D. Hierarchical 3D multiphysics modelling in the design and optimisation of SOFC system components. Int J Hydrogen Energy 2011;36:4400–8.
[32] Chapter 8 – Vibrations in fluid–structure interaction systems. In: Flow-induced Vib.; 2014. p. 359–401. Available from: https://doi.org/10.1016/B978-0-08-098347-9.00008-4.
[33] Wick T. Coupling fluid–structure interaction with phase-field fracture. J Comput Phys 2016;327:67–96. Available from: https://doi.org/10.1016/j.jcp.2016.09.024.
[34] Paidoussis MP. Fluid-structure interactions: slender structures and axial flow, vol. 2; n.d.
[35] Belostosky AM, Akimov PA, Kaytukov TB, Afanasyeva IN, Usmanov AR, Scherbina SV, et al. About finite element analysis of fluid – structure interaction problems. Procedia Eng 2014;91:37–42. Available from: https://doi.org/10.1016/j.proeng.2014.12.008.
[36] Rasheed A, Holdahl R, Åkervik E. A comprehensive simulation methodology for fluid-structure interaction of offshore wind turbines. Energy Procedia 2014;53:135–45. Available from: https://doi.org/10.1016/j.egypro.2014.07.222.
[37] Hsu M-C, Bazilevs Y. Fluid-structure interaction modeling of wind turbines: simulating the full machine. Comput Mech 2012;50:821–33. Available from: https://doi.org/10.1007/s00466-012-0772-0.

[38] Khor CY, Abdullah MZ, Lau C-S, Azid IA. Recent fluid–structure interaction modeling challenges in IC encapsulation – a review. Microelectron Reliab 2014;54:1511–26. Available from: https://doi.org/10.1016/j.microrel.2014.03.012.

[39] Trivedi C, Cervantes MJ. Fluid-structure interactions in Francis turbines: a perspective review. Renew Sustain Energy Rev 2017;68:87–101. Available from: https://doi.org/10.1016/j.rser.2016.09.121.

[40] Yang K, Sun P, Wang L, Xu J, Zhang L. Modeling and simulations for fluid and rotating structure interactions. Comput Methods Appl Mech Eng 2016;311:788–814. Available from: https://doi.org/10.1016/j.cma.2016.09.020.

[41] Yeoh GH, Cheung CP, Tu J, Yeoh GH, Cheung CP, Tu J. Chapter 2 – Computational multiphase fluid dynamics framework. In: Multiph. Flow Anal. Using Popul. Balanc. Model. 2014; 17–67. Available from: https://doi.org/10.1016/B978-0-08-098229-8.00002-4.

[42] Brennen CE. Fundamentals of multiphase flow 2013. Available from: https://doi.org/10.1017/CBO9780511807169.

[43] Mahady K, Afkhami S, Kondic L. A volume of fluid method for simulating fluid/fluid interfaces in contact with solid boundaries. J Comput Phys 2015. Available from: https://doi.org/10.1016/j.jcp.2015.03.051.

[44] Roenby J, Larsen BE, Bredmose H, Jasak H. A new volume-of-fluid method in openfoam. In: VII Int Conf Comput Methods Mar Eng Mar 2017; 2017.

[45] Hirt CW, Nichols BD. Volume of Fluid (VOF) methods for the dynamics of free boundaries. J Comput Phys 1981. Available from: https://doi.org/10.1016/0021-9991(81)90145-5.

Chapter 6

# Thermomechanical Modelling of Materials and Components

## Chapter Outline



The dimensional and mechanical stability of materials and components is of paramount importance to their use in everyday life, where they have beensubjected to thermal loading. Everybody knows if concrete roads were made of one continuous piece, cracks would appear, owing to expansion and contraction brought about by the difference between summer and winter temperatures.



Multiphysics Modelling. DOI: https://doi.org/10.1016/B978-0-12-811824-5.00006-7

Do you know why the banging sound in trains is associated with the railroad? This is a simple example of the *thermal expansion* compensation. The tracks are assembled such that space is left between the ends of two track sections on purpose. The gap allows the track to expand when it is heated and contract when it is cooled. Therefore the sound of train wheels rattling is louder in the winter when the gaps are larger and less in the summer.

This difference in expansion can also lead to problems when interpreting the petrol gauge of a car. The actual amount (mass) of petrol left in the tank when the gauge hits 'empty' is a lot less in the summer than in the winter. The petrol has the same volume as it does in the winter when the fuel light appears, but because the petrol has expanded, there is less mass. This is the reason why a driver will probably run out of petrol quicker in the summer season.

Although expansion can be troublesome, it often proves very useful; like the force of contraction when hot metal cools is also utilised in riveting together the steel plates and girders. Ceramics are fired so as to consolidate their final structure. This extendable list of traits makes the thermomechanically induced phenomena very interesting. Especially, the multiphysics scientists and engineers in research and development are strongly encouraged to read, understand and educate themselves in the area. Their contribution to the design and development of various technologies will benefit through meaningful analyses.

As the current chapter focuses on the mechanical response of the materials and components subjected to thermal environments, attention has been given to supply a comprehensive but practical understanding of some thermomechanics fundamentals. It has been aimed to support and improve the understanding of the reader about the thermal–mechanical interactions, as the topic is of great interest to many multiphysics engineers. The reader should notice that the previous chapters built up a foundation on some of the most common fields that multiphysics concerns and the ongoing focal point has been to combine and utilise these knowledge in approaching complex problems. Complementary examples have been used for this purpose.

## 6.1 Thermomechanical Modelling Foundations

The increased interest to this advanced topic dates already back to the discussions published in the early 1950s where mathematical solutions to thermomechanically induced deformation and stresses, comprising various engineering fields have been depicted. Analytical and numerical solutions for the thermoelasticity or thermoplasticity problems of linear and nonlinear nature have been sought after since then [1–15].

The term thermomechanics can be defined as the field of mechanics, studying the relationship between thermal and external loads applied on a body or surface and the intensity of internal forces acting within that body. The subject also concerns with the change in dimensions (length or volume) of the specimen as a function of temperature, as well as deformations. But obviously there is more to it than meets the eye!

The well known term *strain* from basic mechanics courses could basically be defined as a measure of elongation or contraction, causing change in volume and shape of a body. Within the subject of the thermomechanics, this change of the body is due to the thermal effects. The governing equations of thermomechanics are based on those of the theory of elasticity, including the equations of motion, the compatibility equations and the constitutive law. Likewise, problems where the plasticity is considered, the same equations of equilibrium and compatibility as in the theory of elasticity are used in conjunction with the proper constitutive laws and the stress–strain relations.

The effect of the temperature field in the governing equations is integrated through the constitutive law. Actually, one can imagine the theory of linear thermoelasticity as being based on the linear addition of thermal strains to the mechanical strains. And in situations where the plasticity needs to be considered, the equilibrium and compatibility equations remain the same as for elasticity problems. The main difference lies in the constitutive law. In closely coupled thermomechanics problems, the time rate of change of the first invariant of the strain tensor in the first law of thermodynamics is also taken into account. This secures the dependence between the temperature and strain fields, thus creating the coupling between strain and thermal fields.

In many engineering applications it has been sufficient to apply the one-way coupling of the thermal and mechanical problem, as the thermal field affects the stress–strain field but usually not the way back. The temperature distribution that results due to the thermal or thermofluid flow conditions is first determined and the thermal field result is then used to calculate the response of the solid body to the imposed thermal gradients and other applied forces. Even in complex materials such as utilised in geomechanics, this kind of semicoupled approach has shown satisfactory engineering results [16,17]. Of course, if the thermofluid flow affects the geometrical shape due to pressure or thermal effects over time, then the new flow field will be changed, thus the thermal and the thermomechanical behaviour will be different. In situations like this, a close coupling and a two-way fluid structural analysis should be attempted.

As the constitutive law is the distinguishing critical component compared to sole mechanically loaded problems, it would be convenient to shed some light on this aspect to relate it to thermally induced strain and stresses without getting deep in to the mathematical nature of the equilibrium and compatibility equations. The main benefit will be to use the space intended for this chapter effectively and give the reader some information that has not been covered very often in text books.

In a component or material subjected to both thermal and mechanical loads, the total physical strain comprises a mechanical strain and a thermal strain component, both mathematically expressed as a vector. The mechanical strain part contains an elastic, plastic or an additional creep strain term that would arise due to external forces and/or temperature field changes, body forces and exerts normal and shear components. On the other hand, the thermal strain is the measure of the elongation–contraction of the material due to change in temperature. The thermally induced deformations (thermal strain component of the strain energy density of a solid) can be described using the index notation of the thermal strain tensor expressed as:

$$\varepsilon_{ij}{}^{T} = \alpha \Delta T \delta_{ij} \tag{6.1}$$

where $i$ and $j$ are subscripts with values 1, 2, 3 that represent the Cartesian components of the thermal strain vector. Here, $\delta_{ij}$ denotes to the Kronecker delta symbol; $\Delta T = T - T_{ref}$ refers to the difference of the current local temperature to the reference temperature assumed to be at zero initial strain. $\alpha$is the linear thermal expansion coefficient of the material that is expressed as:

$$\alpha = \frac{1}{T - T_{ref}} \int_{T_{ref}}^{T} \alpha'(\vartheta) d\vartheta \tag{6.2}$$

Here, $\alpha'$ refers to the tangential thermal expansion coefficient. The Kronecker delta implies that in the thermal expansion tensor the function is 1 if the variables $i$ and $j$ are equal, otherwise 0. For example,$\delta_{12} = \delta_{21} = \delta_{31} = \delta_{13} = 0$, whereas $\delta_{22} = \delta_{11} = \delta_{33} = 1$. One can imagine this as if $\delta_{ij}$ are components of an identity tensor or an identity matrix of $3 \times 3$.

In an unconstraint state, a homogeneous linear elastic isotropic material with no surface loads and kept at a uniform temperature would freely expand and produce no stresses. For the isotropic behaviour, $\Delta T$ would have the same effect in all directions, thus no stresses would arise. For an anisotropic case, $\alpha$ would be dependent of direction, thus should be used in Eq. (6.1) also as a tensor such as $\varepsilon_{ij}{}^{T} = \alpha_{ij}{}^{T} \Delta T$and the Kronecker delta could be removed.

When the same component is constraint or made of different materials, or if the temperature is changing in the part then thermomechanically induced stresses will ensue. Thus, the 3D thermomechanical problem, comprising the field variables displacement

$ui(x_1, x_2, x_3)$, strain $\varepsilon_{ij}(x_1, x_2, x_3)$and stress $\sigma_{ij}(x_1, x_2, x_3)$ needs to be formulated and solved.

As an example for the mathematical formulation of such a problem, the governing equations for the coupled thermoelastic problem can be expressed as:

$$\underbrace{\sigma_{ij,j} = \rho \ddot{u}_i}_{\text{Equation of motion}} \quad ; \quad \underbrace{\varepsilon_{ij} = \frac{1}{2}\left(u_{i,j} + u_{j,i}\right)}_{\text{Strain-displacement relations}} \quad ;$$

$$\underbrace{\sigma_{ij} = 2\mu\varepsilon_{ij} + \lambda\varepsilon_{kk}\delta_{ij} - (3\lambda + 2\mu)\alpha(T - T_0)\delta_{ij}}_{\text{Constitutive law for a linear homogeneous isotropic thermoelastic material}} \tag{6.3}$$

combining these equations will yield to the description of the equation of motion in terms of displacements expressed as:

$$\mu u_{i,kk} + (\lambda + \mu)u_{k,ki} - (3\lambda + 2\mu)\alpha T_{,i} = \rho \ddot{u}_i \tag{6.4}$$

Eq. (6.4) refers together with the energy equation given as:

$$\underbrace{kT_{,ii} - \rho c\dot{T} - \alpha T_0(3\lambda + 2\mu)\dot{\varepsilon}_{ii} = -R}_{\text{Energy equation}} \tag{6.5}$$

to the displacement–temperature equations of the coupled thermoelasticity for a solid elastic body. Here, $R$ is the heat per unit time per unit volume and $\lambda$, $\mu$ refer to the Lamé constants that are related to the Young's modulus and the Poisson's ratio through:

$$\lambda = \frac{E\nu}{(1 + \nu)(1 - 2\nu)}; \quad \mu = \frac{E}{(1 + \nu)} \tag{6.6}$$

The thermal boundary conditions for the problem are satisfied through either of the following equations:

$$T = Ts; \partial D \quad \text{for } t > t_0 \tag{6.7}$$

$$T,_n + aT = b \text{ on } \partial D \quad \text{for } t > t_0 \tag{6.8}$$

where $a$ and $b$ are either constants or given functions on the boundary. $D$ is the so called Biot's free energy that is based on the definition of entropy flux ($Si$) given as:

$$D = \frac{1}{2}\frac{d}{dt}\int_v \frac{T_0}{k} S_i S_i \, dV \tag{6.9}$$

These conditions are based on the applied temperature, convection or radiation on the boundary. The mechanical boundary conditions can be specified through the traction vector on the component boundary. The traction components are related to the stress tensor through the so called Cauchy's formula considered as:

$$t_i^n = \sigma_{ij} n_j \quad \partial D \quad \text{for } t > t_0 \tag{6.10}$$

where $t_i^n$ refers to the traction components (the force vector on a cross section divided by that cross-section's area) on the boundary whose outer normal vector is $n$. Using the constitutive law for linear thermoelasticity along with the strain–displacement relations, the traction components are then presented in terms of the displacement components as:

$$t_i^n = \mu\left(u_{i,j} + u_{j,i}\right)n_j + \lambda u_{k,k} n_i - (3\lambda + 2\mu)\alpha(T - T_0)n_i \tag{6.11}$$

Both Eqs (6.10) and (6.11) relate the traction boundary conditions to the stress or displacement fields [18]. The following example in Fig. 6.1 should demonstrate the practical calculation of the traction components on a simplified sample.

Imagine the object with a cross-sectional area of 300 $mm^2$ that is pulled in tension by a 3000 N force (*red*) in the $x$-direction. So the (arbitrarily chosen) rightward pointing internal force vector (*blue*) is $3000i$ N. The traction vector would then be $\left(1/300 \text{ mm}^2\right)3000i \text{ N} = 10i$ MPa.

The units of MPa rather than N are used, but it is a vector, not a stress tensor! If the object is cut on one side with an angle of

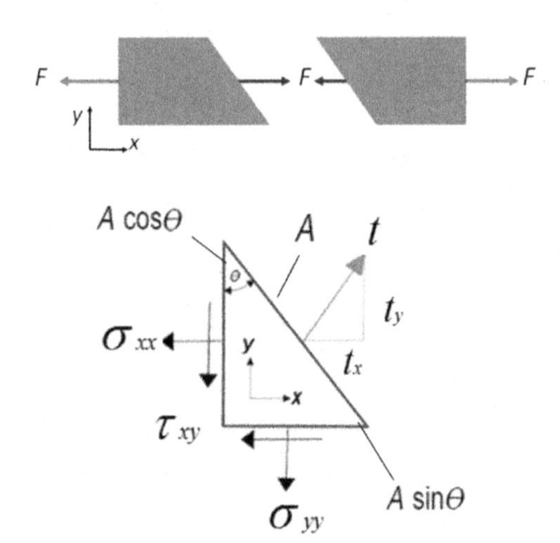


**Figure 6.1** The relationship between the traction vector and stress state.

30 degrees, the effective applicable are would be 346 mm$^2$ and the traction vector would be calculated as:

$$t = \left(\frac{1}{346\ \text{mm}^2}\right) 3000i\ \text{N} = 8.67i\ \text{MPa}$$

Considering $n$ as the unit normal vector to the surface, and $s$ for the unit vector parallel to it, means that the normal stress and shear stresses can be calculated as:

$$\sigma = t.n = (8.67, 0, 0).(\cos 30°, \sin 30°, 0) = 7.5\ \text{MPa}$$
$$\tau = t.s = (8.67, 0, 0).(-\sin 30°, \cos 30°, 0) = -4.33\ \text{MPa}$$

$$\sigma_{xx} A\cos\theta + \tau_{xy} A\sin\theta = t_x A$$
$$\tau_{xy} A\cos\theta + \sigma_{yy} A\cos\theta = t_y A$$

setting the sum of forces on the sample object cancels the area $A$ in the equation and considering that the $\cos\theta$ and $\sin\theta$ are only the components of the unit normal to the surface ($n = (\cos\theta, \sin\theta)$) that the vector is acting, it is useful to replace $\cos\theta$ and $\sin\theta$ with $n_x$ and $n_y$ and for the extended 3D case as $n_z$, leading to:

$$\begin{aligned} &\sigma_{xx}n_x + \tau_{xy}n_y + \tau_{xz}n_z = t_x \\ &\tau_{xy}n_x + \sigma_{yy}n_y + \tau_{yz}n_z = t_y \\ &\text{or for the 3D} \\ &\tau_{zx}n_x + \tau_{zy}n_y + \sigma_{zz}n_z = t_z \end{aligned} \tag{6.12}$$

The stress tensor in Eq. (6.10) would result in nine separate stress components. For example, if we consider the following stress tensor given as:

$$\sigma = \begin{bmatrix} 50 & 10 & 30 \\ 10 & 95 & 20 \\ 30 & 20 & 15 \end{bmatrix}$$

if the traction vector on the surface with a normal vector of let's say $n = (0.40, 0.60, 0.693)$ is calculated, the mathematical formulation would be expressed in matrix form as:

$$\begin{Bmatrix} Tx \\ Ty \\ Tz \end{Bmatrix} = \begin{bmatrix} 50 & 10 & 30 \\ 10 & 95 & 20 \\ 30 & 20 & 15 \end{bmatrix} \begin{Bmatrix} 0.40 \\ 0.60 \\ 0.93 \end{Bmatrix} = \begin{Bmatrix} 46.8 \\ 74.9 \\ 34.4 \end{Bmatrix}$$

thus, the resulting traction vector would be:

$$t = 46.8i + 74.9j + 34.4k \text{ MPa}$$

with the used area of say 346 mm$^2$, the force applied could be calculated as:

$$F = 16192.8i + 25915.4j + 11902.4k \text{ MPa}$$

Likewise, it is possible to predict the normal and shear stresses, acting on the same surface as:

$$\sigma = \begin{Bmatrix} 0.40 & 0.60 & 0.693 \end{Bmatrix} \begin{bmatrix} 50 & 10 & 30 \\ 10 & 95 & 20 \\ 30 & 20 & 15 \end{bmatrix} \begin{Bmatrix} 0.40 \\ 0.60 \\ 0.93 \end{Bmatrix} = 87 \text{ MPa}$$

To calculate the shear stress on the surface, let's assume to choose a surface in the direction s given as:

$$s = (-0.83, 0.55, 0.0)$$

If it is desired to choose another surface parallel to the actual surface and perpendicular to the first direction $s$, the unit normal vector has to be crossed with the first tangential vector, prior to calculating the shear stress.

Hence,

$$\begin{aligned} n \times s &= (0.40i + 0.60j + 0.693k) \times (-0.83i + 0.55j + 0.0k) \\ &= -0.384i - 0.576j + 0.721k \end{aligned}$$

now the shear stress can be calculated as:

$$\begin{aligned} \tau &= \{ -0.384 \quad -0.576 \quad 0.721 \} \begin{bmatrix} 50 & 10 & 30 \\ 10 & 95 & 20 \\ 30 & 20 & 15 \end{bmatrix} \begin{Bmatrix} 0.40 \\ 0.60 \\ 0.93 \end{Bmatrix} \\ &= -36 \text{ MPa} \end{aligned}$$

## 6.2 Thermomechanical Material Properties

To simulate the thermomechanical behaviour of materials and components requires reliable values for both thermal and the mechanical properties of the materials. The quality of the predicted results depend on the accurate prediction of the thermal field. Therefore considering the thermophysical processes the component is subjected to and the material data utilised in the modelling is very important. Since material properties are temperature dependent, the actual local temperature experienced by the component will alter the physical properties used and influence the overall calculations.

In general, the thermal properties of the materials describe the response to change in thermal energy, i.e. heat. The atoms in the materials are in vibration, which constitute the thermal energy of the material. Thermal properties of the materials like the coefficient of thermal expansion, thermal conductivity or the heat capacity are all occulted phenomena of this vibration.

The mechanical response of the thermal and external loads is determined by proper use of the mechanical properties such as Young's modulus and the Poisson's ratio. Here, the Young's modulus takes higher values and is an indication of the stiffness of a material and is also highly temperature dependent. Hence, it is also important to consider its temperature dependency during the simulations. Most of the metals show Poisson's ratio values within a range of 0.25–0.35.

Table 6.1 presents some properties that can be used in simulations. These data have been gathered and utilised during scientific studies in energy materials research. As they are not covered in common text books, it will be of benefit for the multiphysics analyst.

## 6.3 Model Idealisation

The used computational domain for the solution of the thermomechanical problem in materials and components is very important, as it affects the accuracy and reliability of the predicted results. In Chapter 1, Introduction to Multiphysics Modelling, attention has been drawn to the effect of load applications and constraints. To

**TABLE 6.1** Some Material Properties for Analyses

| Property | Anode | Cathode | Electrolyte | Interconnector | Sealant |
|---|---|---|---|---|---|
| Specific heat capacity (J/kg-K) | 595 | 573 | 606 | 502 | 458 (FeCrAlY) |
| Thermal conductivity (W/m-K) | 5.84 | 1.86 | 2.16 | 22.1 | 186 (FeCrAlY) |
| Poisson's ratio (–) | 0.32 | 0.36 | 0.313 | 0.3 | 0.33 (FeCrAlY) |

complement the given information, the current section will focus on the effect of the assumed geometrical model and its practical limits. Moreover, the effect of simplifications on the analysis is discussed. The following example shows a specimen that has been subjected to a tensile test under constant temperature. The sample has been subjected to an applied force of 1500 N. After a full analysis, due to the symmetry of the sample, the half section of it has been simulated using finite element method (FEM). The third sample is a quarter of the same sample that has also been simulated and compared. The results demonstrate that there is a good agreement among all three results as the Von Mises stress distribution of the simulated samples show (Fig. 6.2).

The important thing the reader needs to consider whilst performing those kind of geometrical simplification is to apply appropriate boundary conditions to the cut regions. Thus in the half section of the geometry, the face where the model has been cut needs to be defined as symmetry, whereas in the quarter section, an additional symmetry condition has to be set to the bottom face where the

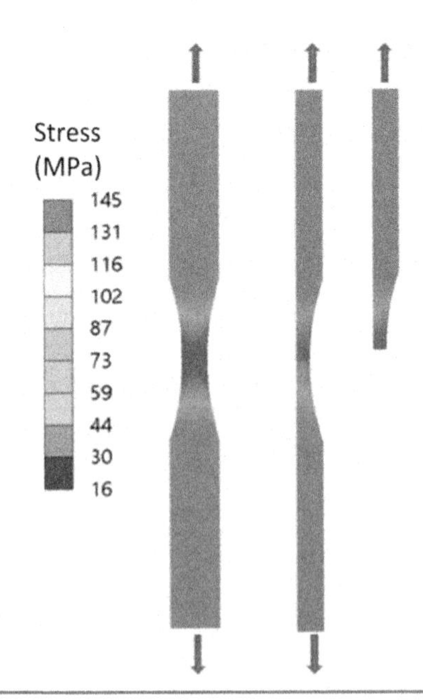


**Figure 6.2** Effect of geometrical simplification on the thermomechanically induced stresses.

model has been cut. The predictions of the analyses show that a geometry simplification for this problem is valid. The results show numerical maximum results of 145–146 MPa, thus minor differences are present.

The second point the analyst needs to take care of is the load that has been applied. This needs to be halved as well, in order to be equivalent to the full model results, otherwise in the linear analysis the resulting stress yields to the double value showing approximately 290 MPa! As demonstrated, if the loads and geometry show completely symmetry, geometrical simplifications of this kind may be used and is certainly of advantage, regarding computational efficiency. Concerning the question whether to employ a 2D or 3D model, the answer would be to use a 2D model, if the geometry and loads of the problem can be completely depicted in one single plane, otherwise a 3D analysis is appropriate. Engineers perform usually the geometrical approximations for preliminary investigations of processes or the design of the structure is proposed by employing a trial section of the whole component.

The loads that may act on the component are estimated and the mathematical models usually tested. The detailed geometry needs to be used in the design and detailing stage where the constituents are designed to satisfy safety and long-time service requirements. This is also valid for the thermomechnical analysis, thus the characteristics of the physical component must be represented as accurate as possible by the used model. Fig. 6.3 depicts another example, where the importance of the careful consideration of simplification and geometrical assumptions has been stressed.

The model of a fuel cell section, comprising the physical wire mesh and metal interconnector plate has been compared to a hypothetical wire mesh structure, modelled as a continuum and has been detailed in Ref. [19]. A uniformly distributed mass of the solid body subjected to a specified temperature (the section is of cm size and within a region of constant temperature inside the whole component) is considered, using the porosity of the continua. The differences in the thermomechanical behaviour have been investigated.

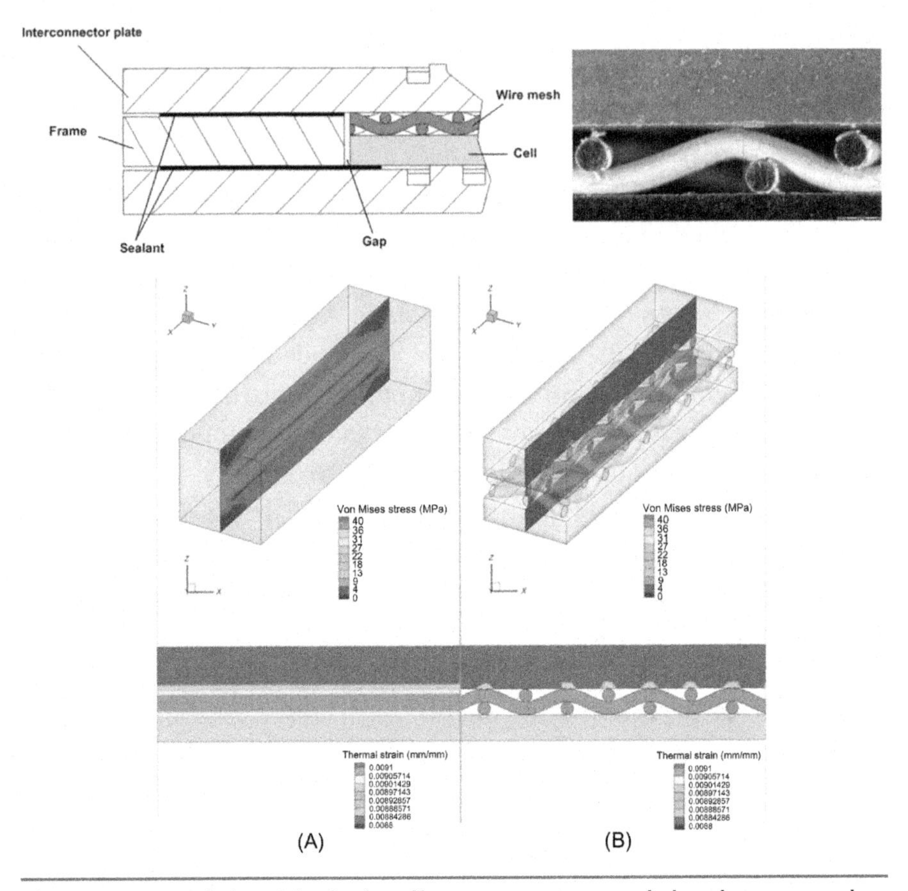


**Figure 6.3** Solid oxide fuel cell cross section and the thermomechanically induced stress, thermal strain plots: the continuum approach (A) showed locally discrepancies compared to the resolved geometry (B) that led to practical importance.

The used model is a section of the given fuel cell layer that has been depicted in Fig. 6.3.

The most important practical output from the presented stress and thermal strain distributions is to shed light on the approach differences. The effect of both the approaches as well as the 2D versus 3D attempts show clearly the loss of information; particularly, the local contact region of the wavy structure to the neighbour components. In practice, the knowledge of this information has been invaluable, as certain effects could not be captured with the continuum approach. The thermomechanically induced stresses at this region affects the electrical contact of the components, which could be comprehended better. Moreover, the illustrated 2D planes show clearly that these kind of physical components need to be investigated using 3D thermomechanical models. Hence, the decision of which approach to choose must be carefully planned.

## 6.4 Practical Review Example

So far, the subject thermomechanical modelling has been handled from a theoretical point of view, as well as some critical aspects that multiphysics analysts need to draw attention to. The current section targets to support the topic, considering a practical problem.

### 6.4.1 Example: Thermomechanical Analysis of an Electrically Heated Component

A simple experimental set up has been used to investigate the thermomechanical response of an metal alloy component. For this purpose, electrically heated brass cartridges have been used to heat-up the component from both sides. The heating up continues until the component reaches a maximum steady temperature. It is assumed that the main component is made of metal alloy and some aluminium alloy frame components are present in the interior section. The *assumed* thermomechanical properties are given in Table 6.2. Note

**TABLE 6.2** Thermomechanical Material Properties

| Assumed Properties | Metal Alloy | Aluminium Alloy | Brass |
|---|---|---|---|
| Thermal conductivity (W/m-K) | 23 | 114 | 123 |
| Young's modulus (MPa) | 50,000 | 71,000 | 100,000 |
| Poisson's ratio (–) | 0.29 | 0.33 | 0.30 |
| Coefficient of thermal expansion (1/K) | $1.528E^{-05}$ | $2.3E^{-05}$ | $1.9E^{-05}$ |

that the cartridges are made of brass; however, as they are not joint with the component and the heat transfer over time does not play a role for this example, the properties of it do not play a major role.

Assume that the component is exposed to ambient air of 22°C with a heat transfer coefficient of 5 W/m$^2$-°C. The cartridge has a constant temperature of 150°C. The problem is illustrated together with the applied boundary conditions in Fig. 6.4.

1. Considering the given conditions, determine the temperature distribution of the component.
2. Determine the distributions of the thermal strain and the induced thermomechanical stress of the component after it has reached a steady state temperature distribution.

To solve the problem, the first thing is to understand the physical state of the problem, what the targets are and how to approach it. The problem considered convective and conductive heat transfer. No fluid flow inside the component is present, mitigating a comprehensive computational fluid dynamics analysis. Thus the thermal field will affect the deformation behaviour of the component, but not vice versa. The FEM can be used to solve for the thermally loaded structural analysis.

The solution utilises a hexahedral dominating numerical grid of 27,897 elements, satisfying mesh-independent result. Both cartridges are designated with temperature boundary conditions with a constant temperature of 150°C. The material type has been assigned to the component as to register the properties. The remaining surfaces

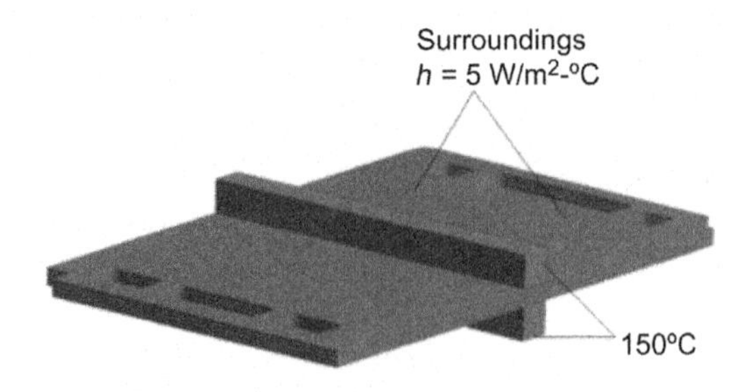


**Figure 6.4** Demonstration of the electrically heated component.

subject to the environment have been defined as convection boundary conditions with the assigned heat transfer coefficient of 5 $W/m^2$-°C and the bulk temperature of 22°C, as given in the problem description. Fig. 6.5 illustrates the numerical grid and the temperature distribution of the prediction.

The temperature contours show the symmetrical distribution of the thermal field on both the left and right side of the cartridges. As there are no additional heat sources or a heat sink, the maximum region is as expected at the vicinity of the cartridges. The heat is conducted smoothly in all direction to the component. It shows that the geometrical and load symmetry is enabled, thus giving the hint to the analyst that for process investigations the use of half of the body can be used (mind that a fourfold symmetry is not given as geometrical hollow regions are not the same!). The convection effect is noticed as the thermal field is decreasing in the lateral direction along the component.

As the current problem is interested on the structural response of the thermally loaded component in the steady state, the transient terms were not considered. Therefore the thermal mass does not play an important role here. The structural response of the component can be calculated either using coupled elements having the displacement as degree of freedom or structural elements need to be defined meaning the numerical grid must be set using structural elements. As mechanical loads are not used, the only load utilised is the thermal distribution.

Fig. 6.6 shows in brief the results of the simulation.

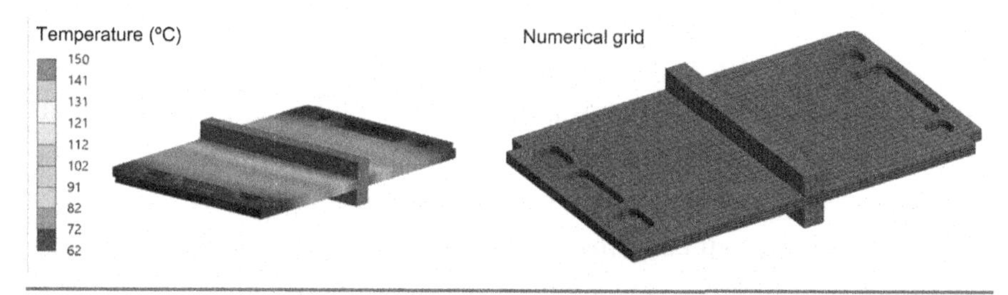


**Figure 6.5** Demonstration of the predicted temperature distribution and the utilised numerical grid of the problem.

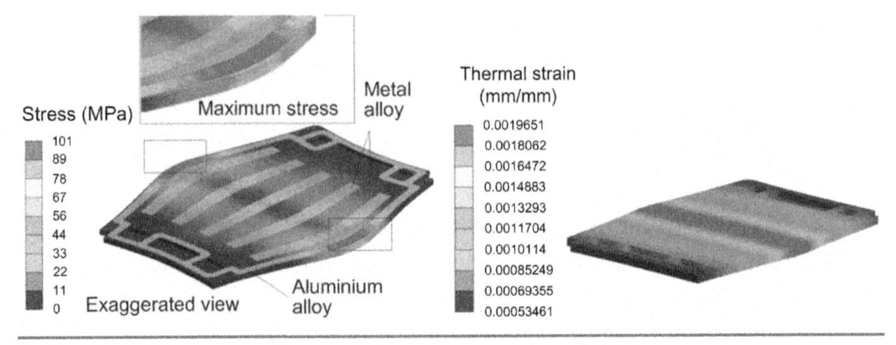


**Figure 6.6** Stress and thermal strain results of the component.

As the metal components are of concern, the analyst may evaluate the results using the Von Mises stress theory. The stress contour plot results (equivalent stress shows no negative sign as the principal stresses come out of the root sign!) reveal that the component has been subjected to three-dimensional deformation as it has been well demonstrated using an exaggerated scale view. The blade components inside are positioned lateral, bonded to the main component. Thus this constraint results in stresses. However, the largest displacement is observed on the hottest region on the main component. As the thermal strain plot assists the analyst in understanding the phenomena, it can be seen that the end section where the cartridges are located try to expand the most. However, the stiffer aluminium material in this example resists more to deflection compared to the softer metal alloy. As these are bonded, stresses arise and reach the maximum at this vicinity.

After the initial simulations, the analyst can focus on precision. Therefore, to take advantage of the symmetry, a computational less expensive half section can be used. To reach quantitative accuracy, temperature-dependent material properties and nonlinear material behaviour can be implemented, etc. The review sample demonstrates how an analyst decides how to approach, define and seeking for a solution in a thermomechanical problem. Moreover, methods to evaluate and explain the relation between the results and the reasons need to be chosen.

# References


[1] Noda N. Thermal stresses. FL: CRC Press, Boca Raton; 2002.

[2] Nowacki W. Dynamic problems of thermoelasticity. Netherlands: Springer; 1975.

[3] Sih GC. Thermomechanics of solids: nonequilibrium and irreversibility. Theor Appl Fract Mech 1988;9:175–98. Available from: https://doi.org/10.1016/0167-8442(88)90030-4.

[4] Wetton RE. Thermomechanical methods; 1998. p. 363–99. Available from: https://doi.org/10.1016/S1573-4374(98)80009-5.

[5] Wong AK. Thermoelastic stress analysis. Ref Modul Mater Sci Mater Eng 2016. Available from: https://doi.org/10.1016/B978-0-12-803581-8.03343-9.

[6] Bechtel SE, Lowe RL, Bechtel SE, Lowe RL. Chapter 4 – The fundamental laws of thermomechanics. In: Fundam. Contin. Mech. 2015. 115–54. Available from: https://doi.org/10.1016/B978-0-12-394600-3.00004-6.

[7] Bechtel SE, Lowe RL, Bechtel SE, Lowe RL. Chapter 5 – Constitutive modeling in mechanics and thermomechanics. In: Fundam. Contin. Mech.; 2015. 157–80. Available from: https://doi.org/10.1016/B978-0-12-394600-3.00005-8.

[8] Boley BA, Weiner JH. Theory of thermal stresses. Mineola: Dover Publications; 2012.

[9] Carrera E, Fazzolari FA, Cinefra M, Carrera E, Fazzolari FA, Cinefra M. Chapter 11 – Static and dynamic responses of coupled thermoelastic problems. In: Therm. Stress Anal. Beams, Plates Shells; 2017. 345–60. Available from: https://doi.org/10.1016/B978-0-12-420066-1.00014-8.

[10] Carrera E, Fazzolari FA, Cinefra M, Carrera E, Fazzolari FA, Cinefra M. Chapter 7 – Computational methods for thermal stress analysis. In: Therm. Stress Anal. Beams, Plates Shells; 2017. 241–90. Available from: https://doi.org/10.1016/B978-0-12-420066-1.00009-4.

[11] Dhondt G. The finite element method for three-dimensional thermomechanical applications. New York: John Wiley & Sons; 2004.

[12] Kleiber M, Kowalczyk P. Introduction to nonlinear thermomechanics of solids. Cham: Springer International Publishing; 2016.

[13] James J. Chapter 7 – Thermomechanical analysis and its applications. In: Therm. Rheol. Meas. Tech. Nanomater. Charact; 2017. 159–71. Available from: https://doi.org/10.1016/B978-0-323-46139-9.00007-4.

[14] Lu Y, Zhang X, Xiang P, Dong D. Analysis of thermal temperature fields and thermal stress under steady temperature field of diesel engine piston. Appl Therm Eng 2017;113:796–812. Available from: https://doi.org/10.1016/j.applthermaleng.2016.11.070.

[15] Peksen M. Numerical thermomechanical modelling of solid oxide fuel cells. Prog Energy Combust Sci 2015;48:1–20. Available from: https://doi.org/10.1016/j.pecs.2014.12.001.

[16] Booker JR, Carter JP, Small JC, Brown PT, Poulos HG. Some recent applications of numerical methods to geotechnical analysis. Comput Struct 1989;31:81–92. Available from: https://doi.org/10.1016/0045-7949(89)90170-3.

[17] Carter JP, Booker JR. Finite element analysis of coupled thermoelasticity. Comput Struct 1989;31:73–80. Available from: https://doi.org/10.1016/0045-7949(89)90169-7.

[18] Hetnarski RB, Eslami MR. Thermal stresses – advanced theory and applications. Netherlands: Springer; 2008.

[19] Peksen M. A coupled 3D thermofluid-thermomechanical analysis of a planar type production scale SOFC stack. Int J Hydrogen Energy 2011;36:11914–28.

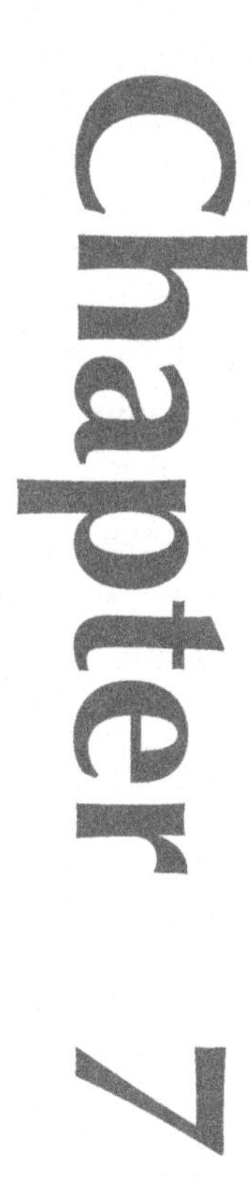


# Multiphysics Modelling of High-Performance Materials

## Chapter Outline



In our daily life, we are confronted by a multitude of different material-related phenomena whose essential role often extends over many scales in time and space. Almost every aspect of technology comprises natural or artificial materials. Buildings we live in [1–6], vehicle tires [7], several automotive components [8–10], fibres and textiles [11–21], machine components [22–26] are all composed of materials. Thus modern engineering materials play a major role in



Multiphysics Modelling. DOI: https://doi.org/10.1016/B978-0-12-811824-5.00007-9

technological designs. Tissue and bones [27] or fuel cells [28–30] and much more that we now take for granted, all comprise high-tech materials.

Many of today's environmental and technological challenges require various methods for determining and probing the properties of a vast range of materials. Advances in material modelling promote and assist these attempts. With increasing advances in the accuracy, realism, and predictive capabilities, modern multiphysics modelling has become an essential part of the research and product development efforts.

In any case, be it the foundation of living matter or the engineering design of products and processes, the underlying phenomena of modelling of high-performance materials span a huge and hierarchically organised sequence of time and length scales. To describe and comprehend such multiscale and multiphysics phenomena through predictions and advanced numerical methods is a challenging task.

The modelling of high-performance materials is an essential process to predict processes accurately. Beyond the purely monetary value, it profoundly affects the reliability of components and processes, whose functionality relies upon in various technological fields of science and engineering. While on electron level processes can occur within seconds or minutes, a metallic creep deformation proceeds over thousands of hours; likewise corrosive effects for instance may proceed over decades. Hence, different time scales are present.

On the other hand, high-performance materials modelling comprises an assortment of existing materials. Most of the readers are aware of typical solid materials such as metals, polymers and composites that have been utilised in various modelling activities, or they are familiar with natural materials such as soils and rocks used in geotechnical studies [31–36]. Ultimately, the structural or thermofluid flow response of such materials subjected to a given load (mechanical or thermal) in a defined process has constituted the bulk of multiphysics analyses available in the literature.

The provided approaches and predictions are comprehensive and provide accurate results for most classical materials that are subjected to strain paths of reasonable complexity. However, the continuing demand for technological advances and understanding of some classes of complex materials and processes has posed new multiphysics modelling challenges, urging scientists to seek for new solutions. Thus, in recent years, the numerical simulation of high-performance materials has been incorporated into the procedures of many manufacturing processes.

Industry is seeking not only to estimate loads and energy requirements with higher accuracy but also for a better understanding of their products. Thus estimation for possible product defects and life is also expected. This automatically influences the attempted modelling strategy. The predicted behaviour of interest, i.e. structural, thermal or flow, as well as the numerical techniques and input data used to solve the problem become important when addressing these type of demands.

The current chapter aims to give a brief insight to the practical modelling of different high-performance materials, regardless of their applications. It should be noted that it is not possible to cover every conceivable type of material or combinations. However, the reader should comprehend the importance of mimicking and implementation of the appropriate material model in the practical multiphysics situations.

## 7.1 Polymeric Materials

Modelling and simulation of high-performance materials is a massive research field, still open to new developments and strategies. One of those areas are the porous materials. Most of the readers are familiar with the wide range of functional and structural applications of porous materials with favourable thermal, acoustic and energy-absorption properties.

Various concepts on modelling the thermal, fluid flow or structural behaviour of porous materials are well known. Particularly, many multiphysics analysts are already accustomed with the theory of porous media [37–46] and the representative volume element [47–50] applications in continuum mechanics [34,51,52] and try to utilise where appropriate.

The reader may refer to many of the informative text books for detailed theoretical foundations of porous media, as well as their modelling in general [31,33,35,53–67]. The following example, however, is a sporadic one that has been chosen on purpose to summarise and combine the simulation of a porous media made of polymer materials to predict its complex melting behaviour during a thermal bonding processing. The mathematical modelling of the microstructural and rheological properties of the used fibres is depicted.

Processes related to melting and solidification indeed encompass a large range of engineering and scientific disciplines, including agriculture, artificial freezing of ground, construction, food processing and thermal energy storage; therefore its modelling is a challenging practice for the multiphysics analyst. The introduced example analysis is a demonstration of a multiphysics model that was used as part of a product and machinery optimisation study [68,69].

The problem is described such that heated convective air is passed through an unbonded permeable polymeric web structure, usually wrapped around a perforated drum. Due to forced convection, all the transported fusible continuously laid fibres, soften, melt and flow towards the crossover point of adjacent fibres in contact. A glue-like bond forms to provide strength to the polymer web while leaving the system (Fig. 7.1).

Multiphysics modelling is used to predict the complex melting behaviour of the web, over time. Properties such as fibre thickness, melting point and latent heat of fusion will be considered. The liquid fraction of the polymeric material will be observed. In practice this gives an indication about the maturity level of the bond between fibres. A continuum formulation is adopted, thus separate phase conservation equations are not required. The used model equations are based on volumetric averaging of the microscopic conservation equations and occur over a temperature range that considers the solidus and liquidus temperatures of the used material.

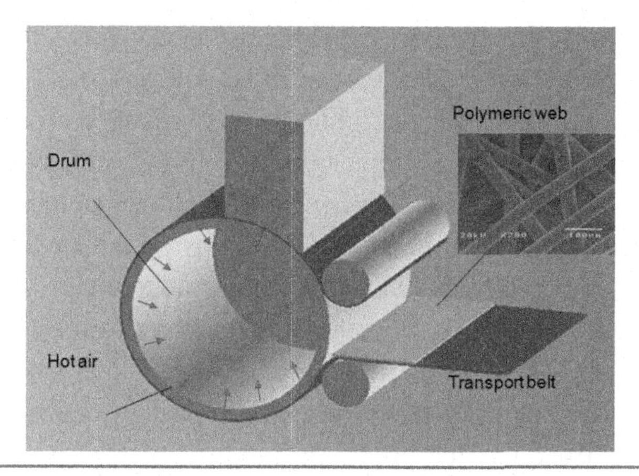


**Figure 7.1** Porous polymeric web production system.

The assumption is justified because unlike pure substances, multi-constituent systems do not exhibit a sharp interface between solid and liquid phases. Discrete phase change rarely occurs in practice. Since the complicated phase interfaces are not required to be monitored, a single zone formulation is suitable for treating the continuous transition between the solid and the liquid phases. Hence, the use of macroscopic model equations is based on volumetric averaging of the microscopic conservation equations.

An enthalpy-based phase change model is applied [70,71]. For the phase change, the enthalpy of the material is assumed as the sum of the sensible heat of the liquid–solid mixture, in addition to the contribution of the latent heat. The modelled phase change is assumed to occur linearly with latent heat as a linear function of temperature. To consider the structural properties of the polymeric web, the individual properties of the polymeric fibre layers are integrated within the model. Note that for demonstration purposes individual information about the materials are not detailed. The reader may refer to the given references for a detailed realistic industrial analysis.

User-defined scalars and the diffusivity, permeability of the porous web are implemented with macros. This enables the analyst to link the microscale and macroscale information. Thus the thermofluid process information from the system level could be coupled with the individual material behaviour and component structure (porosity).

Fig. 7.2 illustrates the physical bicomponent fibres of which the porous web is made of. It is the outer sheath region of these fibres that softens and bonds with neighbouring fibres as to form the web. The inner core layer has a higher melting point and gives the composite fibre the structural stability. The contact points forming a bonded region can be noticed for a setting of 20 μm of the SEM micrograph. The bicomponent fibres have a diameter of 31.71 μm and a sheath fraction is 0.273 μm. The sheath material has a melting temperature of 221°C.

To model this melting behaviour, the liquid fraction of the polymeric sheath fibre ($\gamma$) has to be defined. The porosity of the web with 0.9 mm thickness is not considered as constant. It is considered as a function of the representative latent heat content of each computational cell. The liquid fraction of the sheath fibre ($\gamma$) is in the molten region defined as completely liquid, so that it is equal to the sheath volume fraction ($\varepsilon$). This yields to $\varepsilon(1-\varphi)$ of the air volume fraction (porosity) $\varphi$ of the volume and is 1. So, the latent heat tends to approach to its maximum when the porosity is considered to be one.

In the solid region, $\varepsilon = \gamma = 0$. Close to the melting temperature, the volume element might be simultaneously in liquid and solid states. In this situation the liquid fraction is considered to be between $0 < \gamma < 1$ and $0 < \gamma < \varepsilon(1-\varphi)$. When the polymeric material reaches its phase melting temperature, the porosity is considered to be 0.5.

Accordingly, the energy equation for the matrix-solid–liquid mixture can be mathematically defined as:

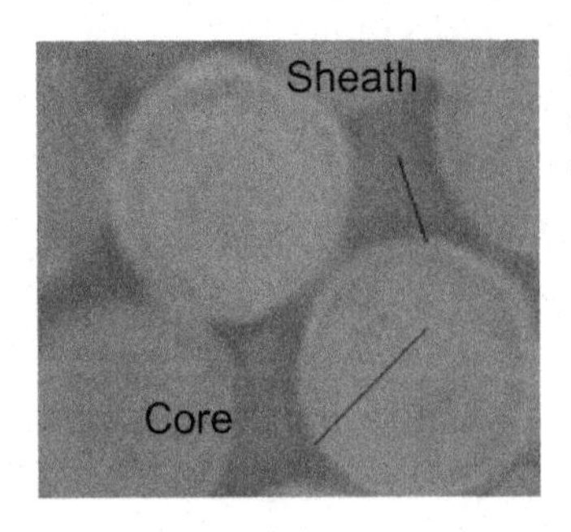


**Figure 7.2** SEM photomicrograph of sheath-core type bicomponent fibre inside the web.

$$\frac{\partial}{\partial t}\left[\varphi\rho_a \boldsymbol{h_a} + (1-\varphi)\left(\varepsilon\left(\gamma\rho_l \boldsymbol{h_l} + (1-\gamma)\rho_s \boldsymbol{h_s}\right) + (1-\varepsilon)\rho_c \boldsymbol{h_c}\right)\right] + \nabla.(\rho_a \boldsymbol{h_a}\vec{\boldsymbol{u}}) = \nabla.(\boldsymbol{k_{eff}}\nabla \boldsymbol{T}) \tag{7.1}$$

$h\zeta$ is defined as the enthalpy of the constituents $\zeta$; with $a$ referring to air; $l$ is the liquid state, $s$ the solid state of the sheath fibre material and $c$ represents the core part of the bicomponent fibre that is retained in solid state due to its thermophysical properties. $k_{eff}$ and $T$ are the effective thermal conductivity and temperature of the composite mixture, respectively.

Despite that the phase change is assumed to occur at a discrete temperature, in a volume element that contains the porous matrix, as well as the solid and liquid states, the average temperature may be slightly higher or lower than the melting temperature. Therefore it is assumed that both solid and the liquid phases may exist simultaneously in a volume element, if its temperature is within a small temperature range $\Delta T$, on either side of the fusion temperature.

When a volume element is subjected to phase change (i.e. $0<\gamma<1$), the change in the mean fluid enthalpy is because of the change in the sensible heat of the liquid–solid mixture, as well as the latent heat contribution. This can be formulated as:

$$d\left[\gamma\rho_l h_l + (1-\gamma)\rho_s h_s\right] = \left[\gamma\rho_l c_l + (1-\gamma)\rho_s c_s\right]dT + \rho_l \Delta h d\gamma \tag{7.2}$$

where $c_\zeta$ is the specific heat capacity of the given constituent $\zeta$. Substituting Eq. (7.2) into Eq. (7.1) leads after simplification to the conservation equation of energy that considers the melting of the liquid saturated porous media.

$$\overline{\rho c}\frac{\partial T}{\partial t} + \rho_a c_a(\overline{u}.\nabla T) = \nabla.(k_{eff}\nabla T) - (1-\varphi)\varepsilon\rho_l \Delta h\frac{\partial\gamma}{\partial t} \tag{7.3}$$

The mean thermal capacitance ($\overline{\boldsymbol{\rho c}}$) of the solid–liquid mixture is expressed as:

$$\overline{\boldsymbol{\rho c}} = \varphi\rho_a \boldsymbol{c_a} + (1-\varphi)\left[\varepsilon\left(\gamma\rho_l \boldsymbol{c_l} + (1-\gamma)\rho_s \boldsymbol{c_s}\right) + (1-\varepsilon)\rho_c \boldsymbol{c_c}\right] \tag{7.4}$$

The effective thermal conductivity can be calculated using the theory of mixtures. Considering the sheath and core fractions of the fibres, it can be obtained from:

$$k_{eff} = \varphi k_a + (1 - \varphi)[\varepsilon(\gamma k_l + (1 - \gamma)k_s) + (1 - \varepsilon)k_c] \tag{7.5}$$

The employed computational multiphysics model is depicted in Fig. 7.3. The images illustrate the considered fibres within the web in detail. The used continuum model substitutes the physical computer geometry, where experimentally determined temperature distributions are applied on top and bottom faces over time. This mitigates the use of additional system components and enables to focus on the multiphysics melting process within the web component. The experimentally determined entry velocity of 0.65 m/s has been used.

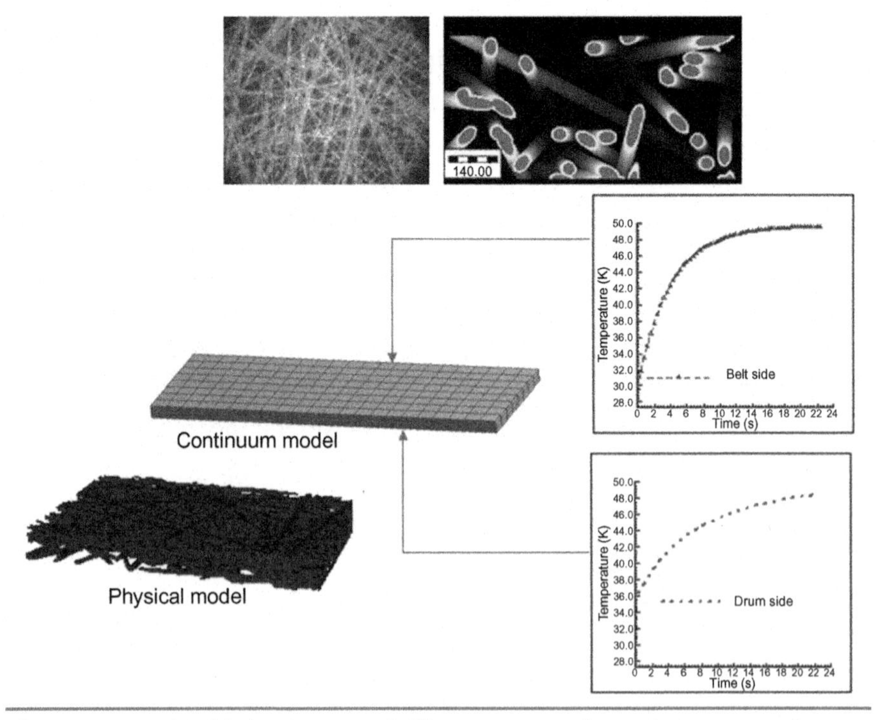


**Figure 7.3** Multiphysics modelling approach: using a continuum model to mitigate geometrical complexities and focus on the melting process.

Fig. 7.4 illustrates the liquid fraction and the temperature distribution from the front view of the structure.

The predicted results show that the liquid fraction is zero at 10 s. The sheath fibre layer is still in solid state at this time. At 15 s the sheath fibre layer of the bicomponent fibre is no more in total solid state. The liquid fraction of the sheath fibre ($\gamma$) and the porosity $\varphi$ lays between 0 and 1. Note that the temperature at this period is slightly approaching the melting temperature of the sheath material assumed as 221°C.

The latent heat is associated with the liquid fraction; thus it is also nonzero. The volume is simultaneously occupied by the liquid and solid phases at the melting temperature. At this temperature the difference between the liquid and solid enthalpies is equal to the latent heat of fusion. The mathematical model assumes that if the mean temperature is within $2\Delta T$, around the fusion temperature, the simultaneous existence of solid and liquid within the volume

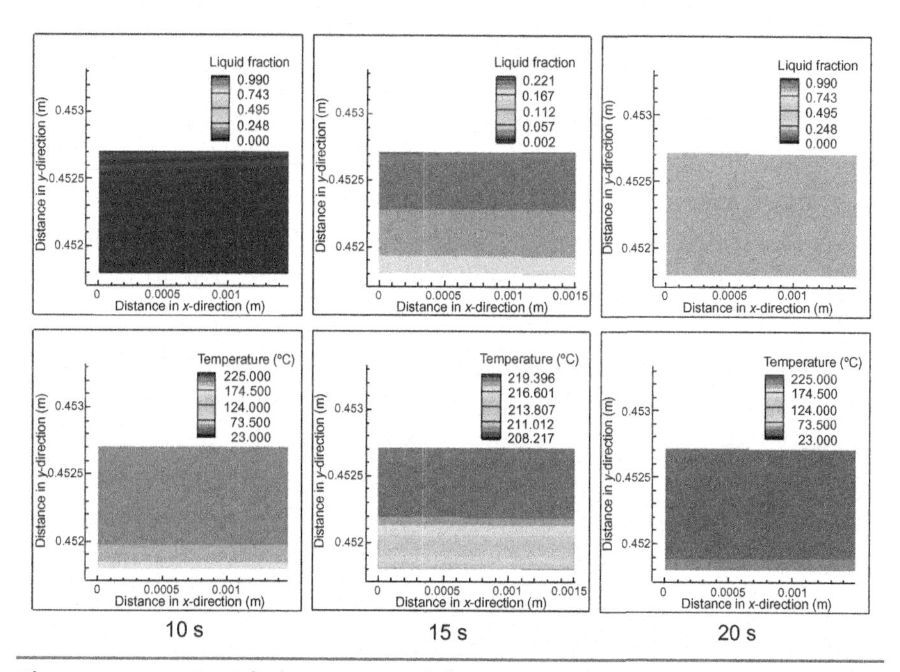


**Figure 7.4** Liquid fraction and temperature contour plots of the porous web component for a time interval of 10–20 s.

may occurs. Another nonlinear model may choose another assumption. The mean temperature is below the fusion temperature, hence the liquid fraction stays below 0.5.

The example aims to shed light on the multiphysics modelling capability to investigate detailed phase change processes during heating of polymeric materials. Attention has been drawn to mathematically account for phase fractions and porosity, as well. Likewise, with suitable mathematical expressions, melting and solidification processes of different materials can be simulated. The approach demonstrates the inclusion of geometrical information such as fibre thickness, sheath fraction, and thermophysical properties like melting temperature, latent heat of fusion and the liquid fraction, mitigating the need for the physical geometry.

## 7.2 Metal Materials

Throughout Chapters 2–6, fundamental processes and various aspects of multiphysics modelling of materials and components have been discussed. Particularly, in Chapter 4, Multiphysics Modelling of Structural Components and Materials, and Chapter 6, Thermomechanical Modelling of Materials and Components, the structural response of solid materials has been handled in detail. The current section demonstrates practical problems comprising different materials; thereby, attracting attention to some technical points.

When materials made of metal are subjected to external forces, their structural response obeys the constitutive laws of elasticity and plasticity. Inelastic or plastic deformation has been shown to occur when the stress is higher than the yield strength in previous topics. The yield criterion and hardening effects have been elucidated. Also, the elastic response of materials and components has been depicted. When modelling high-performance materials, the previously emphasised points are still valid.

The used models, describing the material behaviour are important, as they influence the accuracy of the results directly. Each material has its characteristic behaviour; therefore it is important to account for its behaviour that will be different for a rubber compared to aluminium. The required models, depicting the characteristic stress–strain curve are usually derived from experimental measurements.

The following sample analysis compares the simulation results of a metal material where the stress–strain behaviour of the material is predicted using a bilinear curve and compared to the analysis utilising a multilinear curve representation. The differences of the models are depicted in Fig. 7.5.

As demonstrated, the bilinear model results shown on the top section considers the tangent modulus and the yield strength that predicted a total strain less than the multilinear approach that considers multiple stress–strain points, thus it is visible that the total strain in reality is more than predicted by the bilinear model.

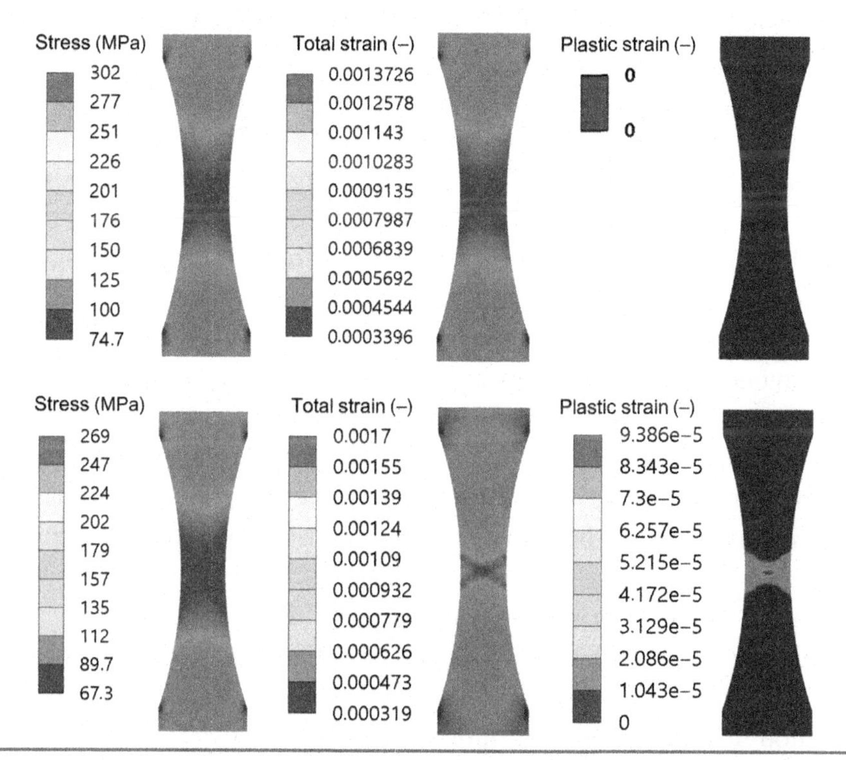


**Figure 7.5** Bilinear model versus multilinear model comparison.

The plastic strain already occurred in the multilinear case, leading to a much higher total strain compared to the bilinear model. The stress distributions show the similar behaviour with the difference that the bilinear model allows for higher stress values, thus no plasticity occurs that explains the lower total strain values.

## 7.3 Hyperelastic Materials

One of the high-performance materials often used in engineering applications are the elastomers. These comprise natural and synthetic rubbers, which are often amorphous. Because of their molecular chains, they become moderately straightened and untwisted under a tensile load. Thus when the load is removed, they revert back to their initial state. As a result of their moderate volume change, elastomers are almost incompressible. Their stress–strain relationship can be highly nonlinear; however, their response is usually assumed to be as isotropic, isothermal and elastic.

The hyperelasticity is not defined as a rate formulation as it is in plasticity considered. Instead, the total-stress versus total-strain relationship is defined through a strain energy potential. The term stretch ratio, relating the length of the specimen after loading to its initial length is used to provide a measure for the deformation. These are defined in principal directions that are needed to determine the strain energy potential (or strain energy function). Through this function, the stress and strain can be calculated.

A practical simplicity of this method is that the analyst can use a curve fitting procedure to translate experimentally determined stress–strain data to convert into strain energy density function coefficients that can be implemented in various hyperelastic mathematical models.

The following example demonstrates the application of two widely used hyperelastic models, i.e. the Neo-Hookean solid model and the Ogden 3$^{rd}$ model. The Neo-Hookean model is a simple thermodynamics-based description of an incompressible hyperelastic material. The Ogden model is a more flexible model, describing fitted experimental data. This has the advantage of directly using test data. Both models are used to solve the nonlinear stress–strain behaviour of materials, undergoing large deformation. For industrial applications, comprising cross-linked polymer chains such as plastics or rubber-like materials they are suitable examples for the reader to utilise for initial studies.

For the demonstration of the models, experimental measurement data obtained from three kinds of tests (Uniaxial tension test, Biaxial tension test and Planar shear test) is utilised in combination to simulate the structural behaviour of chloroprene rubber (neoprene) and rubber specimens (Fig. 7.6).

The models show the typical initial linear section, which plateaus afterwards. It's known that reasonable results are obtained for both models, despite with some limitations to describe particular stress states [72].

In the following example, the mechanical deformation of a shoe sole is simulated using multiphysics modelling. The considered shoe sole is mechanically loaded on the surface contacting the human foot. The predicted total deformation results are compared for the used different materials. Fig. 7.7 illustrates the total deformation of the simulated cases.

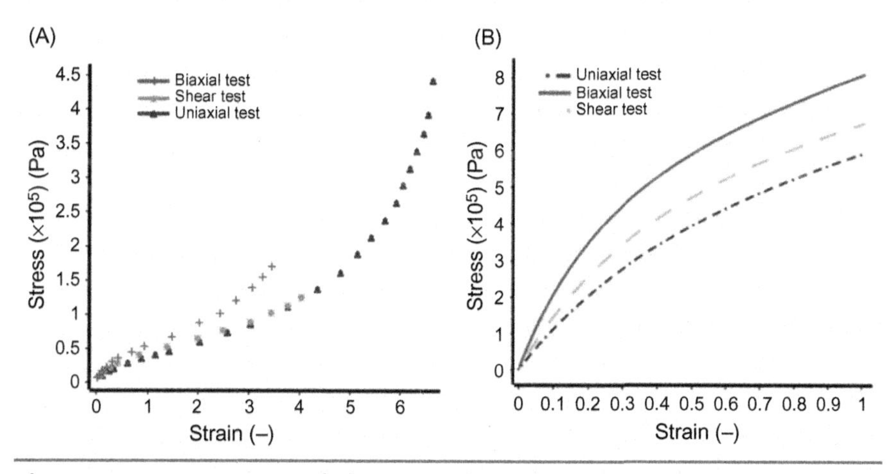


**Figure 7.6** Experimental data sample utilised as a combination: neoprene (A); rubber (B).

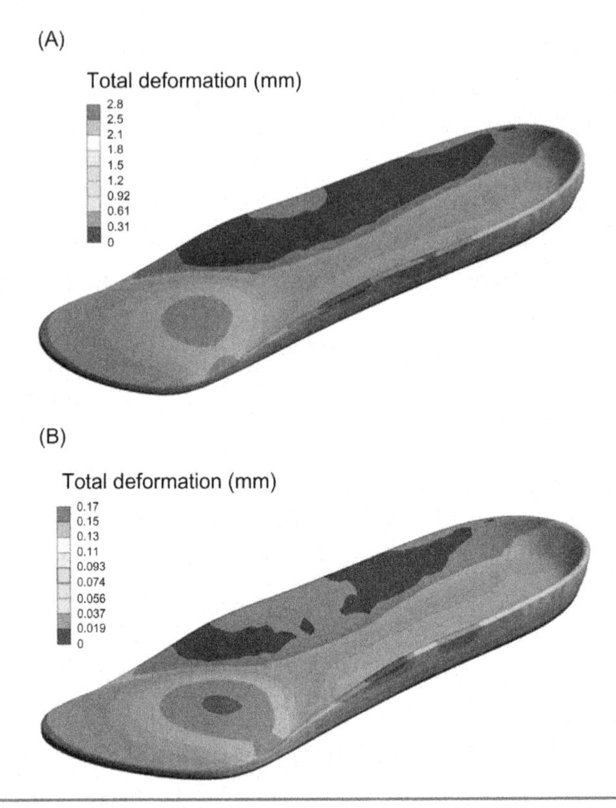


**Figure 7.7** Total deformation of the shoe sole: neoprene (A); rubber (B).

The results reveal that the use of neoprene material deforms with a maximal value of 2.8 mm more than the rubber, reaching approximately 0.17 mm for the same loading. Note that due to the magnitude differences, the results were not set to the same legend scale. Note that for 1:1 comparisons, it is convenient to set the same legend scale.

## 7.4 Composite Materials

From basic courses it is well known that composite materials are mixtures or combinations of two or more materials that are joined together. The combination of multiple materials offers superior properties compared to the material's standalone behaviour. Hence, they result in improved material properties. Usually, one of the materials build the matrix or binding chemical and the other is known as the reinforce material, such as glass-reinforced polyester resin, where the glass fibres increase the strength of the resin. Thus such a combination is favoured in high-tech engineering applications.

Composite materials modelling may utilise macro- or micromodels or sections of a component of interest. The multiphysics simulation may be concerned with the drape behaviour [73–75], or material layers, i.e. at ply or laminate level, as to perform typical structural analyses [13,15–17,21,76]. Thereby, advanced grid elements such as through thickness elements based on shell or solid nature are used to simulate delamination processes that appear due to interlaminar crack between the plies. Many parameters such as ply material type, thickness and angle can be considered. Advanced techniques also utilised in fracture mechanics such as eXtended Finite Element Method (XFEM), J-Integral [77,78], Virtual Crack Closure [79–82] or cohesive zone modelling are often utilised.

Limitations such as the requirements of an initial crack or a very fine mesh around the crack tip may challenge the analyst. Modern meshless methods try to mitigate these challenges, however, facing discontinuity problems or numerical integration issues, thus lots of research still ongoing within this field, as well.

Most of the composite materials are subjected to high loads that lead to failure. There are basically three forms the analyst should be aware of whilst modelling: delamination, matrix cracking and fibre fracture.

The following example aims to demonstrate a simple coupled fatigue crack growth analysis used in engineering structures [83,84]. When an existing crack in the material is of concern, fracture mechanics techniques are usually utilised to determine whether or not the crack will propagate. The energy released by the crack over a certain area is found and the ratio is compared with the critical energy release rate of the material. In practice, if the prediction exceeds the critical energy value, the crack is said to be propagating.

Fig. 7.8 depicts a specimen that depicts the so-called crack opening mode-peeling (Mode I).

Therefore the crack is present on the centreline. Note that if the crack would be off-centre, the crack would involve also Mode II that is concerning in calculation of the energy release for nonsymmetric components. The challenge for the analyst is the combination of all three modes, which requires special attention.

The singularity-based method of the modern XFEM [85–89] is utilised, providing the analyst a convenient technique to investigate the fatigue crack propagation without resorting to model the cracks or remeshing the crack-tip regions as the crack propagates. The Paris's Law [90,91] is chosen for the considered calculation of Mode I and is expressed as:

$$\frac{da}{dn} = C(\Delta K)^m \tag{7.6}$$

where $C$ and $m$ are material constants. For the sake of the study, the Paris's Law constants have been assumed as $C = 1.9\text{E}^{-13}$ and $m = 3$, which need to be determined in practice, experimentally. The

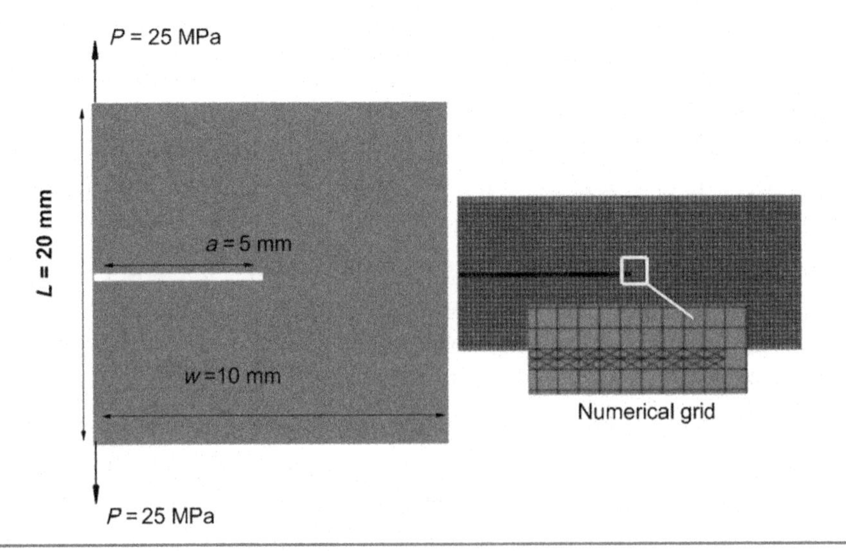


**Figure 7.8** Schematic description of the sample cracked specimen to simulate the crack growth.

crack growth rate *da/dn* is a function of the stress/load ratio, which is chosen as 0, as to mimic a constant amplitude cycling. It is the crack growth per cycle. $\Delta K$ is attributed to the difference of stress intensity factors at the maximum and minimum loads. $a$ refers to the crack length.

The model is chosen for its practical ease. The compression effects and large deflection, as well as plasticity effects are ignored. A linear elastic isotropic approach is considered. An initial crack is already presented. The loading is assumed to be stepwise. The general approach to use a fixed-time stepping is adopted. Crack propagation within an element is assumed to occur along a constant path and direction. The current example has a length to width ratio of 0.5. A cyclic load of $P = 25$ MPa has been applied as a constant amplitude. The material is assumed to have a Young's Modulus of 170 GPa and Poisson's ratio of 0.3. The theoretical prediction of the stress intensity factor can be calculated using the expression:

$$KI = f\left(\frac{a}{w}\right)P\sqrt{\pi a};$$

$$f\left(\frac{a}{w}\right) = 1.119 - 0.231\left(\frac{a}{w}\right) + 10.548\left(\frac{a}{w}\right)^2 - 21.72\left(\frac{a}{w}\right)^3 + 30.387\left(\frac{a}{w}\right)^4 \quad (7.7)$$

Fig. 7.9 illustrates the predicted number of cycles, as well as the difference of the stress intensity factors, occurring at the maximum and minimum loads ($\Delta K$) as a function of the crack extension within the specimen. The results are in good agreement with the theoretically determined calculation.

Fig. 7.10 depicts the stress distribution of the specimen. Notice the high stress value that is due to the linear approach. The reason has been discussed during the linear versus nonlinear approach.

The analyst should be conversant with the detailed processes of crack initiation and propagation, prior switching to more complex multiple crack or scenarios such as surface cracks propagating into a solid.

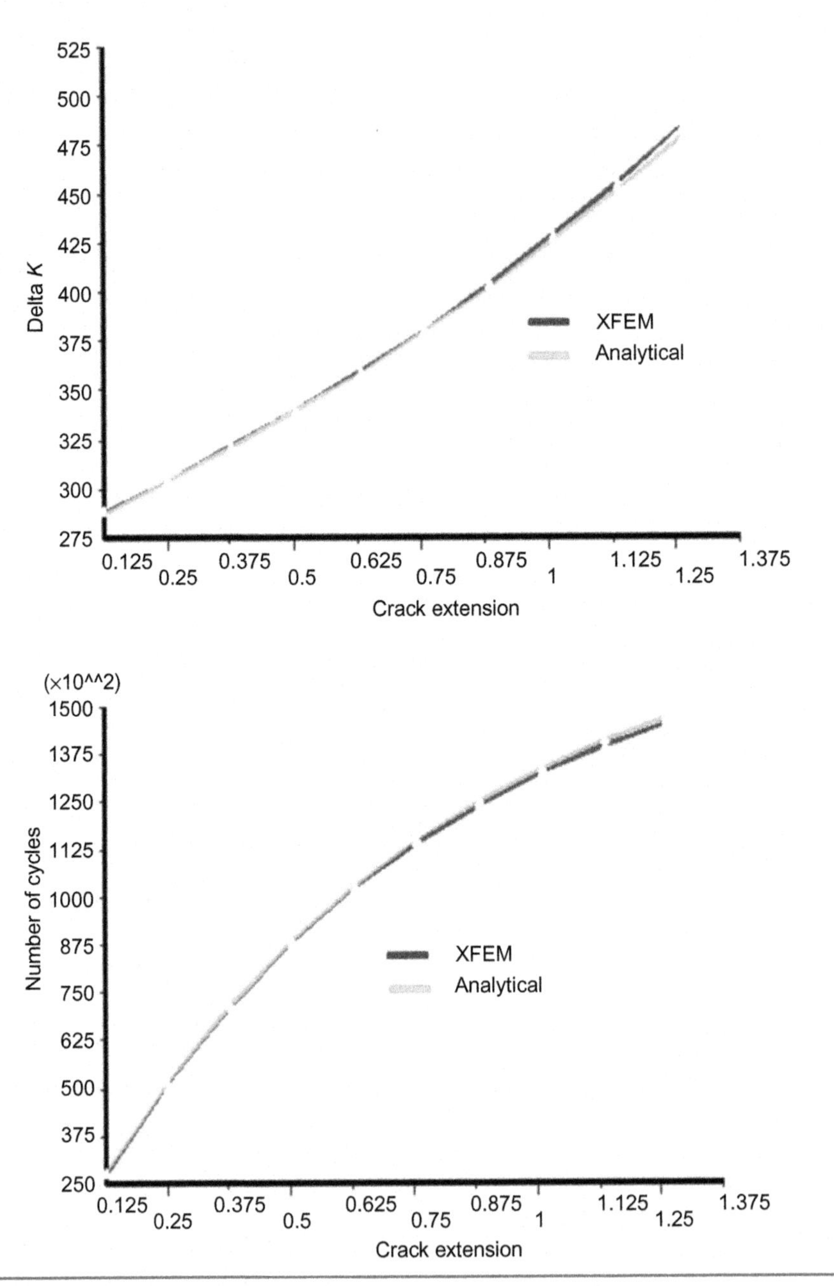


**Figure 7.9** Variation of the stress intensity factors at maximum and minimum loads ($\Delta K$, Delta $K$) with the crack extension.

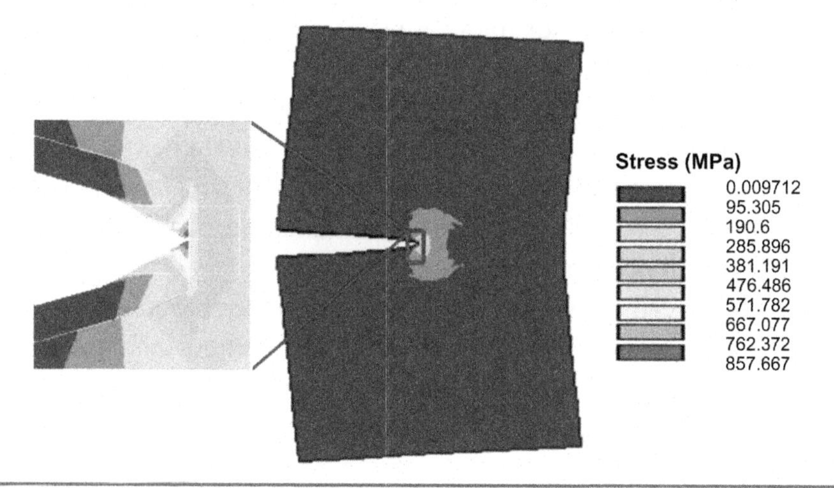


**Figure 7.10** Stress distribution of the specimen after crack growth has occurred.

## 7.5 Ceramic Materials

High-performance materials comprising ceramics are ideally suited for applications where a combination of properties such as wear resistance, hardness, stiffness and corrosion resistance are important. They typically possess high melting points and low conductivities. In addition, engineering ceramics have relatively high mechanical strength at high temperatures. As they often exhibit close thermal expansion coefficients to metals, they are utilised in bonding metal components.

Modelling those materials usually tend to answer questions, regarding the thermomechanical behaviour or the investigation of crack and damage propagation. This is because most of them show brittle nature; therefore, at large loads, rather than accommodating a shape or volume change, materials break. The undeformed material stores elastic energy through stretches and releases this energy by dissipating into the breaking chemical bonds, i.e. fractures.

Typically, the three types of fracture modes, i.e. Mode I, Mode II and Mode III are of interest [92,93]. However, from a multiphysics point of view these kind of analyses are of concern when detailed fracture information is of interest.

Thus the following example aims to complement the example demonstrated in Section 7.4. It has been mentioned that the interpretation of crack growth analyses often make use of the energy release rates. This has been important to determine whether or not the investigated crack will propagate. In very brittle materials such as some ceramics, if a sharp crack or delamination occurs, a singular stress field will occur around the crack tip.

If material failure would be determined solely based on strength analysis, this singularity would exceed the strength and the crack front would propagate through the structure. However, most of the material cracks would not propagate that fast (lower than the speed of sound).

The current example demonstrates the prediction of the energy release rate, typically denoted as $G_c$. For this purpose, a frequently used double layer cantilever beam specimen is used. The specimen has a crack at the free end and is loaded by specifying uniform equal displacement (shown as $F$). The end of the specimen is constrained. Fig. 7.11 depicts the problem.

The load triggers to open the crack region. The displacements at the nodes on the crack surface are then recorded. The required work to close the crack again is obviously the multiplication of the displacement and the force before the crack closes. In practice, this would be the area under the fracture load-deflection curve. It is this

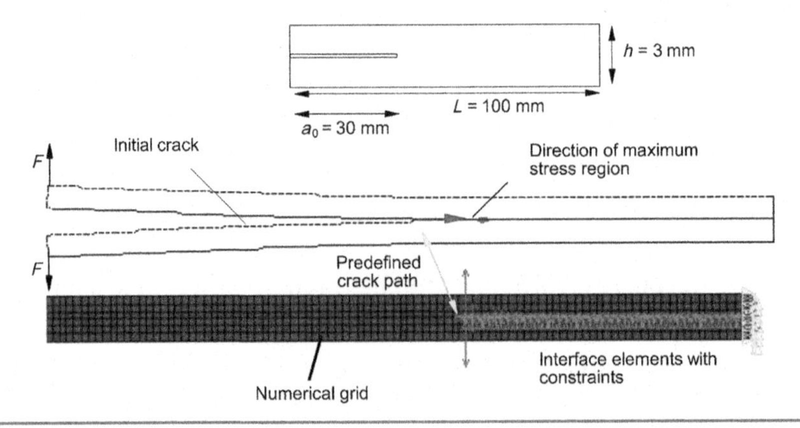


**Figure 7.11** Model description of the double layer cantilever specimen.

work that would be equal to the released amount of energy if the crack opened by one numerical grid node.

If this value would exceed the critical value, the crack would propagate. Again a crack on the centreline approach is used. To determine the amount of crack extension, the summation of the length of the interface elements that are open is measured. For the evaluation, a linear fracture criterion has been assumed. Accordingly, the fracture criterion is assumed to be a function of Modes I, II and III critical energy release rates ($G_I$, $G_{II}$ and $G_{III}$). Fracture occurs when the fracture criterion $f$ exceeds, or is the same value as the fracture criterion ratio $f_c$. The relation is expressed as:

$$f = \left(\frac{G_I}{G_{Ic}}\right) + \left(\frac{G_{II}}{G_{IIc}}\right) + \left(\frac{G_{III}}{G_{IIIc}}\right) \qquad (7.8)$$
$$f \geq = f_c$$

The $f_c$ value here is considered as 1.

The example considers the material as orthotropic, thus the elastic properties are symmetric with respect to the axes. The stiffness properties are assumed to be as $E_{11} = 65$ GPa, $E_{22} = E_{33} = 5$ GPa. The shear modulus is assumed to be as $G_{12} = 2.5$ GPa. The Poisson's ratios have been considered as 0.2 in $V_{12}$ and $V_{13}$ directions, where as in $V_{23}$ a value of 0.4 has been considered. A critical energy release rate of 0.2 has been used for $G_{Ic}$, whereas $G_{IIc}$ and $G_{IIIc}$ have been considered as 0.65.

Fig. 7.12 illustrates the contour plot of the maximum principal stress distribution at the critical region and the variation of the load as a function of displacement.

As theoretically expected, the force increases until the crack starts to grow, which is the point visible in the curve distribution that leads to a peak and denotes the onset of delamination growth. Following this, the reaction force shows a sudden and sharp decrease in magnitude, signalling the beginning of the initial crack growth. This levels down as the crack grows. According to the linear fraction criterion, the fracture will not exceed 1, as the predictions show an energy release rate of maximum 0.171, thus fracture will not occur.

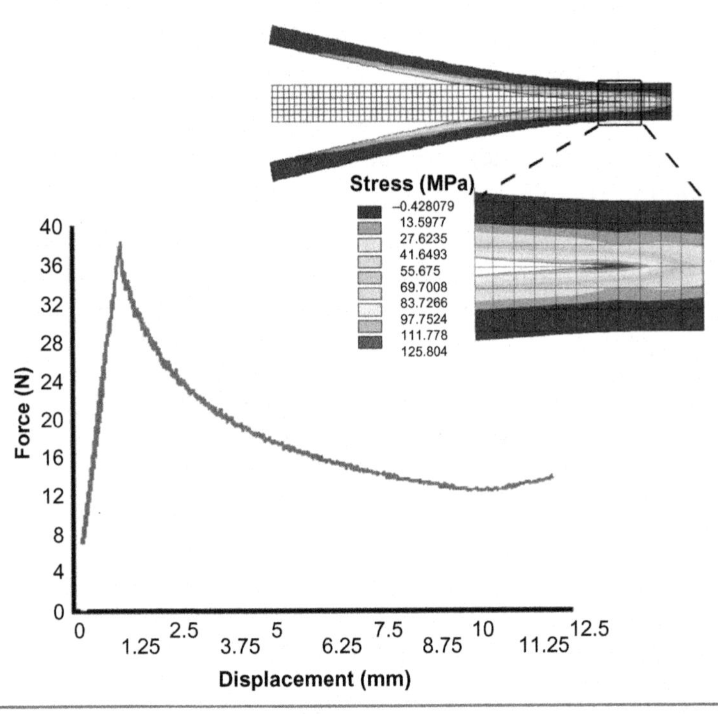


**Figure 7.12** Maximum principal stress distribution of the specimen and the load / displacement response.

The demonstrations in this current chapter showed how effective material modelling of high-performance materials can be used to solve practical issues and push promising cutting edge research. Moreover, it is a fundamental step for describing the behaviour of the used materials in multiphysics processes. In combination with experimental measurements, material modelling can trigger progress, saving time, effort and resources. Their routine use can accelerate product development, increase efficiency and provide fundamental understanding complex processes.

## References

[1] Tang T, Felicelli SD. Micromechanical models for time-dependent multiphysics responses of polymer matrix smart composites. Int J Eng Sci 2015;94:164–80. Available from: https://doi.org/10.1016/j.ijengsci.2015.05.010.

[2] Pan T, Chen C, Yu Q. Three-dimensional micromechanical modeling of concrete degradation under multiphysics fields. Compos Struct 2017;175:7–18. Available from: https://doi.org/10.1016/j.compstruct.2017.05.008.

[3] Lopes CS, Sádaba S, González C, Llorca J, Camanho PP. Physically-sound simulation of low-velocity impact on fiber reinforced laminates. Int J Impact Eng 2016;92:3–17. Available from: https://doi.org/10.1016/j.ijimpeng.2015.05.014.

[4] Gomes CVS, Peixoto DHN, Carrasco EVM, Magalhães MdeC. A novel statistical model for predicting sound absorption in multilayered systems. Energy Procedia 2015;78:1647–52. Available from: https://doi.org/10.1016/j.egypro.2015.11.244.

[5] Løvholt F, Norèn-Cosgriff K, Madshus C, Ellingsen SE. Simulating low frequency sound transmission through walls and windows by a two-way coupled fluid structure interaction model. J Sound Vib 2017;396:203–16. Available from: https://doi.org/10.1016/j.jsv.2017.02.026.

[6] Berardi U, Iannace G. Predicting the sound absorption of natural materials: best-fit inverse laws for the acoustic impedance and the propagation constant. Appl Acoust 2017;115:131–8. Available from: https://doi.org/10.1016/j.apacoust.2016.08.012.

[7] Rafei M, Ghoreishy MHR, Naderi G. Thermo-mechanical coupled finite element simulation of tire cornering characteristics – effect of complex material models and friction law. Math Comput Simul 2017. Available from: https://doi.org/10.1016/j.matcom.2017.05.011.

[8] Osswald TA, Menges G, Osswald TA, Menges G. Rheology of polymer melts. Mater Sci Polym Eng 2012;111–59. Available from: https://doi.org/10.3139/9781569905241.005.

[9] Tanaka T, Nakano R, Tanaka F. Computer simulation for injection and blow molding of polymers. Comput Aided Innov New Mater II 1993;1413–18. Available from: https://doi.org/10.1016/B978-0-444-89778-7.50122-5.

[10] Isayev AI, Kim NH. Co-injection molding of polymers. Inject Molding 2009;851–915. Available from: https://doi.org/10.3139/9783446433731.021.

[11] Polindara C, Waffenschmidt T, Menzel A. A computational framework for modelling damage-induced softening in fibre-reinforced materials – application to balloon angioplasty. Int J Solids Struct 2017;118:235–56. Available from: https://doi.org/10.1016/j.ijsolstr.2017.02.010.

[12] Grabowski K, Zbyrad P, Uhl T, Staszewski WJ, Packo P. Multiscale electro-mechanical modeling of carbon nanotube composites. Comput Mater Sci 2017;135:169–80. Available from: https://doi.org/10.1016/j.commatsci.2017.04.019.

[13] Filipovic N. 3 – Modeling the behavior of smart composite materials. In: Smart Compos. Coatings Membr. 2016; 61–82. Available from: https://doi.org/10.1016/B978-1-78242-283-9.00003-8.

[14] Oskay C. 13 – Multiscale modeling of the response and life prediction of composite materials. In: Numer. Model. Fail. Adv. Compos. Mater. 2015; 351–75. Available from: https://doi.org/10.1016/B978-0-08-100332-9.00013-X.

[15] Melenka GW, Pastore CM, Ko FK, Carey JP. 9 – Advances in 2-D and 3-D braided composite material modeling. In: Handb. Adv. Braided Compos. Mater. 2017; 321–63. Available from: https://doi.org/10.1016/B978-0-08-100369-5.00009-X.

[16] Cerbu C, Botiş M. Numerical modeling of the flax/glass/epoxy hybrid composite materials in bending. Procedia Eng 2017;181:308–15. Available from: https://doi.org/10.1016/j.proeng.2017.02.394.

[17] Abbadi A, Azari Z, Belouettar S, Gilgert J, Freres P. Modelling the fatigue behaviour of composites honeycomb materials (aluminium/aramide fibre core) using four-point bending tests. Int J Fatigue 2010;32:1739–47. Available from: https://doi.org/10.1016/j.ijfatigue.2010.01.005.

[18] Reese S. Meso-macro modelling of fibre-reinforced rubber-like composites exhibiting large elastoplastic deformation. Int J Solids Struct 2003;40:951–80. Available from: https://doi.org/10.1016/S0020-7683(02)00602-9.

[19] Li D, Xia W, Fang Q, Yu W, Shen S. Experimental and numerical investigations on the tensile behavior of 3D random fibrous materials at elevated temperature. Compos Struct 2017;160:292–9. Available from: https://doi.org/10.1016/j.compstruct.2016.10.075.

[20] Kyosev YK. 6 – The finite element method (FEM) and its application to textile technology. In: Simul. Text. Technol.; 2012. 172–222e. Available from: https://doi.org/10.1533/9780857097088.172.

[21] Shokrieh MM, Kamali Shahri SM. 7 – Modeling residual stresses in composite materials. In: Residual Stress. Compos. Mater.; 2014. 173–93. doi:10.1533/9780857098597.1.173.

[22] Kumar N, Yuan W, Mishra RS, Kumar N, Yuan W, Mishra RS. Chapter 6 – Modeling and simulation of friction stir welding of dissimilar alloys and materials. In: Frict. Stir Weld. Dissimilar Alloy. Mater.; 2015. 115–21. Available from: https://doi.org/10.1016/B978-0-12-802418-8.00006-0.

[23] Pang Y, Li Q, Li Q, Luo Q, Chou K-C. Kinetic mechanisms of hydriding and dehydriding reactions in La–Mg–Ni alloys investigated by the modified Chou model. Int J Hydrogen Energy 2016;41:9183–90. Available from: https://doi.org/10.1016/j.ijhydene.2015.11.181.

[24] Clausen AH, Børvik T, Hopperstad OS, Benallal A. Flow and fracture characteristics of aluminium alloy AA5083–H116 as function of strain rate, temperature and triaxiality. Mater Sci Eng A 2004;364:260–72. Available from: https://doi.org/10.1016/j.msea.2003.08.027.

[25] Børvik T, Hopperstad O, Berstad T, Langseth M. A computational model of viscoplasticity and ductile damage for impact and penetration. Eur J Mech – A/Solids 2001;20:685–712. Available from: https://doi.org/10.1016/S0997-7538(01)01157-3.
[26] Chen Y, Clausen AH, Hopperstad OS, Langseth M. Stress–strain behaviour of aluminium alloys at a wide range of strain rates. Int J Solids Struct 2009;46:3825–35. Available from: https://doi.org/10.1016/j.ijsolstr.2009.07.013.
[27] Takabi B, Tai BL. A review of cutting mechanics and modeling techniques for biological materials. Med Eng Phys 2017;45:1–14. Available from: https://doi.org/10.1016/j.medengphy.2017.04.004.
[28] Murakami Y, Matsunaga H. The effect of hydrogen on fatigue properties of steels used for fuel cell system. Int J Fatigue 2006;28:1509–20. Available from: https://doi.org/10.1016/j.ijfatigue.2005.06.059.
[29] Peksen M. Numerical thermomechanical modelling of solid oxide fuel cells. Prog Energy Combust Sci 2015;48:1–20. Available from: https://doi.org/10.1016/j.pecs.2014.12.001.
[30] Abdalla AM, Hossain S, Azad AT, Petra PMI, Begum F, Eriksson SG, et al. Nanomaterials for solid oxide fuel cells: a review. Renew Sustain Energy Rev 2018;82:353–68. Available from: https://doi.org/10.1016/j.rser.2017.09.046.
[31] Barree RD, Conway MW. Beyond beta factors: a complete model for Darcy, Forchheimer, and trans-Forchheimer flow in porous media. Spe 2004;89325. Available from: https://doi.org/10.2523/89325-MS.
[32] Batalha GF. 2.01 – Introduction to materials modeling and characterization. In: Compr. Mater. Process.; 2014. 1–5. Available from: https://doi.org/10.1016/B978-0-08-096532-1.00201-6.
[33] Beavers GS, Sparrow EM. Non-Darcy flow through fibrous porous media. J Appl Mech 1969. Available from: https://doi.org/10.1115/1.3564760.
[34] Bechtel SE, Lowe RL. Fundamentals of continuum mechanics. Elsevier; 2015. Available from: https://doi.org/10.1016/B978-0-12-394600-3.00005-8.
[35] Bernabé Y, Maineult A. 11.02 – Physics of porous media: fluid flow through porous media. In: Treatise Geophys.; 2015. 19–41. Available from: https://doi.org/10.1016/B978-0-444-53802-4.00188-3.
[36] Caicedo B, Ocampo M, Vallejo L. Modelling comminution of granular materials using a linear packing model and Markovian processes. Comput Geotech 2016;80:383–96. Available from: https://doi.org/10.1016/j.compgeo.2016.01.022.
[37] Whitaker S. Flow in porous media III: deformable media. Transp Porous Media 1986;1:127–54. Available from: https://doi.org/10.1007/BF00714689.

[38] Nield DA, Bejan A. Convection in porous media; 2013. Available from: https://doi.org/10.1007/978-1-4614-5541-7.

[39] Bowen RM. Compressible porous media models by use of the theory of mixtures. Int J Eng Sci 1982;20:697–735. Available from: https://doi.org/10.1016/0020-7225(82)90082-9.

[40] Sochi T. Non-Newtonian flow in porous media. Polymer (Guildf) 2010;51:5007–23. Available from: https://doi.org/10.1016/j.polymer.2010.07.047.

[41] Blunt MJ. Flow in porous media – pore-network models and multiphase flow. Curr Opin Colloid Interface Sci 2001;6:197–207. Available from: https://doi.org/10.1016/S1359-0294(01)00084-X.

[42] Carman PC. Flow of gases through porous media. Combust Flame 1957;1:187. Available from: https://doi.org/10.1016/0010-2180(57)90038-X.

[43] Muskat M. The flow of fluids through porous media. J Appl Phys 1937;8:274–82. Available from: https://doi.org/10.1063/1.1710292.

[44] Whitaker S. Flow in porous media I: a theoretical derivation of Darcy's law. Transp Porous Media 1986;1:3–25. Available from: https://doi.org/10.1007/BF01036523.

[45] Yu B. Analysis of flow in fractal porous media. Appl Mech Rev 2008;61:50801. Available from: https://doi.org/10.1115/1.2955849.

[46] Sahimi M, Islam M. Flow and transport in porous media and fractured rocks. DE: John Wiley & Sons, Weinheim; 1996.

[47] Shahzamanian MM, Tadepalli T, Rajendran AM, Hodo WD, Mohan R, Valisetty R, et al. Representative volume element based modeling of cementitious materials. J Eng Mater Technol 2013;136:11007. Available from: https://doi.org/10.1115/1.4025916.

[48] Dong Y, Bhattacharyya D. A simple micromechanical approach to predict mechanical behaviour of polypropylene/organoclay nanocomposites based on representative volume element (RVE). Comput Mater Sci 2010;49:1–8. Available from: https://doi.org/10.1016/j.commatsci.2010.03.049.

[49] Joos J, Ender M, Carraro T, Weber A, Ivers-Tiffée E. Representative volume element size for accurate solid oxide fuel cell cathode reconstructions from focused ion beam tomography data. Electrochim Acta 2012;82:268–76. Available from: https://doi.org/10.1016/j.electacta.2012.04.133.

[50] Li S, Singh CV, Talreja R. A representative volume element based on translational symmetries for FE analysis of cracked laminates with two arrays of cracks. Int J Solids Struct 2009;46:1793–804. Available from: https://doi.org/10.1016/j.ijsolstr.2009.01.009.

[51] Altenbach H. Continuum mechanics., 3. Springer; 2015. p. 354. Available from: https://doi.org/10.1017/CBO9781107415324.004.

[52] Rubin D, Krempl E. Introduction to continuum mechanics; 2010. Available from: https://doi.org/10.1016/B978-0-7506-8560-3.X0001-1.

[53] Sgard F, Castel F, Atalla N. Use of a hybrid adaptive finite element/modal approach to assess the sound absorption of porous materials with meso-heterogeneities. Appl Acoust 2011;72:157–68. Available from: https://doi.org/10.1016/j.apacoust.2010.10.011.

[54] Peksen M, Acar M, Malalasekera W. Optimisation of machine components in thermal fusion bonding process of porous fibrous media: material optimisation for improved product capacity and energy efficiency. Proc Inst Mech Eng Part E J Process Mech Eng 2014;0:1–10. Available from: https://doi.org/10.1177/0954408914545195.

[55] Imomnazarov KK. Modified Darcy laws for conducting porous media. Math Comput Model 2004. Available from: https://doi.org/10.1016/j.mcm.2004.01.001.

[56] Wu Y-S, Wu Y-S. Chapter 10 – Multiphase fluid and heat flow in porous media. Multiph. Fluid Flow Porous Fract. Reserv. 2016;251–64. Available from: https://doi.org/10.1016/B978-0-12-803848-2.00010-6.

[57] Dacun L, Engler T. Modeling and simulation of non-Darcy flow in porous media. In: Proc. SPE/DOE Improv. Oil Recover. Symp.; 2002. Available from: https://doi.org/10.2523/75216-MS.

[58] Satter A, Iqbal GM, Satter A, Iqbal GM. 9 – Fundamentals of fluid flow through porous media. Reserv. Eng 2016;155–69. Available from: https://doi.org/10.1016/B978-0-12-800219-3.00009-7.

[59] Wang YC, Li YD, Wang X. Creep rate induced by surface diffusion of porous media. Phys E Low-Dimensional Syst Nanostructures 2016;75:144–8. Available from: https://doi.org/10.1016/j.physe.2015.09.015.

[60] Tallmadge JA. In: Collins RE, editor. Flow of fluids through porous materials. New York, NY: Reinhold Publishing Co; 1961. p. 270pages.$12.50. AIChE J 1962;8:2–2 . Available from: https://doi.org/10.1002/aic.690080102.

[61] Sobieski W, Zhang Q. Multi-scale modeling of flow resistance in granular porous media. Math Comput Simul 2017;132:159–71. Available from: https://doi.org/10.1016/j.matcom.2016.02.008.

[62] Vossoughi S. Flow of non-Newtonian fluids in porous media. Rheol Ser 1999;8:1183–235. Available from: https://doi.org/10.1016/S0169-3107(99)80017-3.

[63] Chen Z, Lyons SL, Qin G. Derivation of the Forchheimer law via homogenization. Transp Porous Media 2001;. Available from: https://doi.org/10.1023/A:1010749114251.

[64] Huinink H. Two phase flow. Fluids in porous media. Morgan & Claypool Publishers; 2016. p. 9–15. Available from: https://doi.org/10.1088/978-1-6817-4297-7ch9.

[65] Whitaker S. The Forchheimer equation: a theoretical development. Transp Porous Media 1996. Available from: https://doi.org/10.1007/BF00141261.

[66] Huinink H. Single phase flow. Fluids in porous media. Morgan & Claypool Publishers; 2016. p. 6–9. Available from: https://doi.org/10.1088/978-1-6817-4297-7ch6.

[67] Sidiropoulou MG, Moutsopoulos KN, Tsihrintzis VA. Determination of Forchheimer equation coefficients a and b. Hydrol Process 2007. Available from: https://doi.org/10.1002/hyp.6264.

[68] Peksen M, Acar M, Malalasekera W. Computational modelling and experimental validation of the thermal fusion bonding process in porous fibrous media. Proc Inst Mech Eng Part E-J Process Mech Eng 2011;225:173–82.

[69] Peksen M, Acar M, Malalasekera W. Transient computational fluid dynamics modelling of the melting process in thermal bonding of porous fibrous media. Proc Inst Mech Eng Part E J Process Mech Eng 2013;227. Available from: https://doi.org/10.1177/0954408912462184.

[70] Beckermann C, Viskanta R. Natural convection solid/liquid phase change in porous media. Int J Heat Mass Transf 1988;31:35–46. Available from: https://doi.org/10.1016/0017-9310(88)90220-7.

[71] Beckermann C, Viskanta R. An experimental study of solidification of binary mixtures with double-diffusive convection in the liquid. Chem Eng Commun 1989;85:135–56. Available from: https://doi.org/10.1080/00986448908940352.

[72] Kim B, Lee SB, Lee J, Cho S, Park H, Yeom S, et al. A comparison among Neo-Hookean model, Mooney-Rivlin model, and Ogden model for chloroprene rubber. Int J Precis Eng Manuf 2012;13:759–64. Available from: https://doi.org/10.1007/s12541-012-0099-y.

[73] Simon E, Canavan R, Murray P, Geraghty E. Hot drape forming of thermoset matrix composites - characterisation and simulation. In: Proc. 5th Int. Conf. Flow Process. Compos. Mater.; 1999. p. 11–23.

[74] Hübner M, Rocher JE, Allaoui S, Hivet G, Gereke T, Cherif C. Simulation-based investigations on the drape behavior of 3D woven fabrics made of commingled yarns. Int J Mater Form 2016;9:591–9. Available from: https://doi.org/10.1007/s12289-015-1245-8.

[75] Sze KY, Liu XH. Fabric drape simulation by solid-shell finite element method. Finite Elem Anal Des 2007;43:819–38. Available from: https://doi.org/10.1016/j.finel.2007.05.007.

[76] Favata A, Trovalusci P, Masiani R. A multiphysics and multiscale approach for modeling microcracked thermo-elastic materials. Comput Mater Sci 2016;116:22–31. Available from: https://doi.org/10.1016/j.commatsci.2015.10.033.

[77] Begley J, Landes J. The J integral as a fracture criterion. ASTM STP 1972;514:1–23. Available from: https://doi.org/10.1520/STP514-EB.
[78] Zhu XK, Joyce JA. Review of fracture toughness (G, K, J, CTOD, CTOA) testing and standardization. Eng Fract Mech 2012;85:1–46. Available from: https://doi.org/10.1016/j.engfracmech.2012.02.001.
[79] Krueger R. Virtual crack closure technique: history, approach, and applications. Appl Mech Rev 2004;57:109. Available from: https://doi.org/10.1115/1.1595677.
[80] Xie D, Biggers SB. Progressive crack growth analysis using interface element based on the virtual crack closure technique. Finite Elem Anal Des 2006;42:977–84. Available from: https://doi.org/10.1016/j.finel.2006.03.007.
[81] Shivakumar KN, Tan PW, Newman JC. A virtual crack-closure technique for calculating stress intensity factors for cracked three dimensional bodies. Int J Fract 1988;36. Available from: https://doi.org/10.1007/BF00035103.
[82] Leski A. Implementation of the virtual crack closure technique in engineering FE calculations. Finite Elem Anal Des 2007;43:261–8. Available from: https://doi.org/10.1016/j.finel.2006.10.004.
[83] Richard HA, Sander M. Fundamentals of fracture mechanics. Solid Mech Appl 2016;227:55–112. Available from: https://doi.org/10.1007/978-3-319-32534-7_3.
[84] Brocks W. Elastic-plastic fracture mechanics. Solid Mech Appl 2018;244:49–84. Available from: https://doi.org/10.1007/978-3-319-62752-6_5.
[85] Richardson CL, Hegemann J, Sifakis E, Hellrung J, Teran JM. An XFEM method for modeling geometrically elaborate crack propagation in brittle materials. Int J Numer Methods Eng 2011;88:1042–65. Available from: https://doi.org/10.1002/nme.3211.
[86] Pathak H, Singh A, Singh IV, Yadav SK. A simple and efficient XFEM approach for 3-D cracks simulations. Int J Fract 2013;181:189–208. Available from: https://doi.org/10.1007/s10704-013-9835-2.
[87] Datta D. Introduction to eXtended Finite Element (XFEM) Method. arXiv:13085208 [physics.comp-Ph]; 2013.
[88] Jiang Y, Tay TE, Chen L, Sun XS. An edge-based smoothed XFEM for fracture in composite materials. Int J Fract 2013;179:179–99. Available from: https://doi.org/10.1007/s10704-012-9786-z.
[89] Chen L, Rabczuk T, Bordas SPA, Liu GR, Zeng KY, Kerfriden P. Extended finite element method with edge-based strain smoothing (ESm-XFEM) for linear elastic crack growth. Comput Methods Appl Mech Eng 2012;209–212:250–65. Available from: https://doi.org/10.1016/j.cma.2011.08.013.

[90] Nasri K, Zenasni M. Fatigue crack growth simulation in coated materials using X-FEM. C R Méc 2017;345:271–80. Available from: https://doi.org/10.1016/j.crme.2017.02.005.
[91] Paris PC, Gomez M, Anderson W. A rational analytic theory of fatigue. Trend Eng 1961;13:9–14.
[92] Serizawa H, Murakawa H, Lewinsohn CA. Finite element analysis of mode-I & II type fracture behavior on ceramic composite joints. Ceram Eng Sci Proc 2003;24:535–40.
[93] Suresh S, Shih CF, Morrone A, O'Dowd NP. Mixed-mode fracture toughness of ceramic materials. J Am Ceram Soc 1990;73:1257–67. Available from: https://doi.org/10.1111/j.1151-2916.1990.tb05189.x.

Chapter 8

# Multiphysics Modelling of Energy Systems

## Chapter Outline





Multiphysics Modelling. DOI: https://doi.org/10.1016/B978-0-12-811824-5.00008-0

Controlling energy enabled today's human the technical and economical development. Energy with its different forms has been used in communication, technology or various industrial applications. Despite its high percentage of use, the fossil-based energy sources are a problem. They endanger the future generation, because their conversion results through combustion with the release of $CO_2$. Moreover, they will not be available forever. Hence, new primary energy sources like hydro, wind or solar, as well as secondary energy carriers like mechanical and electrical energy became an increased use. As energy systems comprise components that further produce an output, the conversion among those so-called networks results in an interacting complex multiphysics problem (Fig. 8.1). To understand this exchange of detailed information is an important issue, particularly, for the design and development of the technology, as well as building robust systems. This is the reason why multiphysics modelling also gains importance in the field of energy systems.

The current chapter presents the advanced use and application of the interacting multiphysics approaches towards modelling of energy systems-related topics, especially, in light of the major

**Figure 8.1** Network in energy systems.

worldwide challenges confronting energy production, conversion and storage. In general, most of them encompass two or more processes such as the thermofluid flow, species and chemically reactive flow transport or a field of mechanics. However, energy systems are built up of particular components to yield a system. Each component operates different and requires energy sources.

During the utilisation of the resources, different processes arise, resulting in additional complexity. Depending on the technology, additional processes such as electrochemistry, solar tray, chemical reactions or combustion and acoustics may be of concern. As the size, working principle and operating conditions of the concerned energy components vary significantly, the analyst requires a clear understanding of the related processes. Modelling large scale energy systems requires practical experience in advanced simulation techniques. As the technologies and systems within the energy sector are practically countless, some selected topics will be highlighted. It has been targeted to demonstrate the modelling of applications that comprise different processes. These take place in various energy technologies and will be of concern in the near future, as well. Thus worthwhile to consider and round up the chapter.

## 8.1 Thermofluid Flow Processes

Thermofluid flow processes encompass the case where fluid flow and heat transfer have been simultaneously involved. The fundamental chapter related to fluid mechanics dealt with the principles of conservation of mass and momentum, turbulence for the case of fluids in motion. The heat transfer unit comprised the energy transfer. Applications coupling these two fields are considered as thermofluid problems. The following sample chosen for the reader should demonstrate a special application to the so-called spraying process. It has been chosen as it is often observed in energy systems such as in gas turbines and pulverised coal fired boilers.

It is one of the premium class of multiphysics applications in terms of difficulty, as it comprises fluid dynamics and heat transfer interactions, including droplet breakup, fluid force, atomisation and/or evaporation (phase transition), convection, droplet–droplet interactions and droplet–wall interaction. Thus the degree of complexity is very high. Sprays can be dedicated to two-phase fluid flows, which are characterised by a dominating direction of motion. A liquid phase in a discrete form of droplets and ligaments, as well as a gas phase considered as the continuum media concern this kind of problems. The liquid is injected high pressure-driven through an injector into an opening. The jet atomises into tiny liquid fragments and thereafter into droplets. This may evaporate and produce vapour and in some applications can form with air a combustible mixture and ignite. Many factors such as the characterisation of sprays and the assumptions how to treat the phases, their interactions, as well as the effects of turbulence on interface transport rates etc. Hence, it is technically covering lots of details that requires a careful preparation prior deciding on the method, assumptions, etc. [1]. Fig. 8.2 shows an example of the flow distribution during a spraying process.

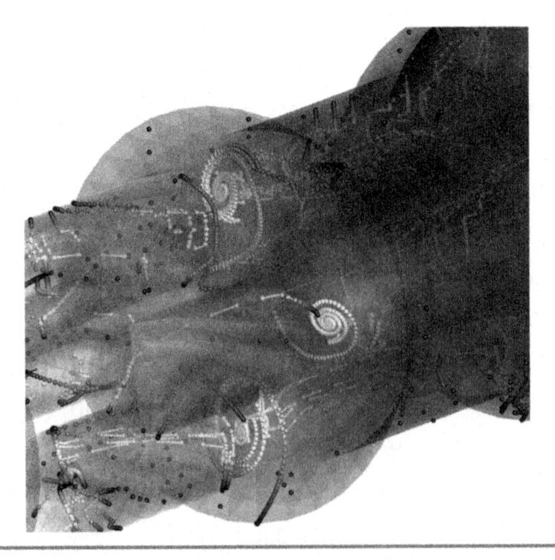

**Figure 8.2** Particle velocities inside a catalyst.

## 8.2 Thermochemical Processes

There are simple and straightforward analogies between electrical, thermal, and fluid systems that are used in 0D modelling of thermal and fluid systems, for example. The analogies between current, heat transfer and fluid flow are intuitive and can be directly applied. Likewise, the analogies between voltage, temperature and pressure are intuitive and useful. However, as the complexity of a system increases the information we gain from those analogies are no more sufficient for a detailed understanding about the systems behaviour. Particularly, chemical systems don't fit well into this analogy scheme.

So far, it has been assumed that whatever enters the domain leaves the domain unchanged. However, in thermochemical cases the species entering the domain will react with each other to form new products and changes not only in time but also in space. This is the reason why multiphysics modelling is also the method to be preferred increasingly in system analyses.

The analysis of reacting systems is usually associated with industrial reactors or being used in chemical engineering. However, a substantial diversity of reacting systems, including biological and environmental, as well as energy systems can be analysed and understood using the same methods. Usually, they are coupled with fluid dynamics and therefore known as chemically reacting species transport or thermochemical flow. Modelling of material or mass transport is necessary if we want to account for mixing or reaction processes such as conversion of biomass to hydrogen, steam reforming or gasification.

The transport of the material can be modelled using either discrete phase modelling [2–4], multiphase modelling [5–17] or by species transport modelling via scalar transport, which will be exemplified in this section. In a multiphysics simulation, including thermochemically reacting species transport, the chemical species transport and chemical reactions include new governing equations that describe the species transport phenomena. However, one single fluid flow field has been solved. The rate of transport of the species is derived from the calculated single fluid flow field. The local species concentrations may affect the flow field itself. As most of the energy applications are subjected to some extent of turbulence, turbulent models are also widely used to describe the flow behaviour. Moreover, the interaction of the chemical reactions with the flow regimes is an important point, as the sensitivity of reaction rates to local changes is complicated by enhanced mixing of turbulent flows.

Chemical reactions or description of the species consist of many different species, thus it becomes very challenging to precisely determine the concentrations and chemical reactions among them. Questions in context to how species mix, react and how fast these happen becomes important. This kind of data is usually determined experimentally and is implemented within the modelling approach. The thermochemical flow problems are generally considered as either homogeneous or heterogeneous. The homogeneous systems consider reactions of species as a single phase. All reactions are assumed to occur in this phase. The gas phase reactions are treated as homogenous systems. Detailed chemistry models are used to model processes such as flame ignition and extinction or calculating pollutants ($NO_x$, CO, etc.).

The heterogeneous reactions comprise the reactions of the reactant and products in different phases. In the case where chemical species are deposited on object surfaces, the treatment is distinctly from the same chemical species in the gas. The rate of deposition is then governed by both chemical kinetics and the diffusion rate from the fluid to the surface. Chemical vapour deposition or catalytic conversion are of this nature.

Regardless of the type of a chemically reacting system, the chemical reacting characteristics require experimental measurements. These are required to include the reaction kinetics, the feeding schedule and the mixing of the species. In a multiphysics analysis, the reaction kinetics is modelled using rate laws, which describe how the reactions take place and how fast they occur. The feeding

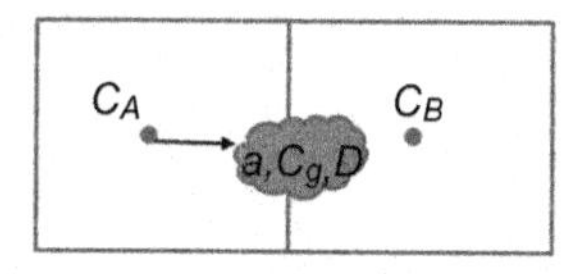


**Figure 8.3** Diffusion flux between two cells.

schedule of the system and mixing of species are important for determining the material balance.

The mixing and transport of chemical species are solved through conservation equations that describe the convection, diffusion and reaction sources for each component species. The local mass fraction of each species is then calculated via the solution of a convection–diffusion equation for the given species that has been elucidated within the subject multiphysics modelling of thermal environments.

In energy systems that encompass thermochemical processes, the material is not only transported through the velocity field (convective term) but also due to the gradients of species concentrations that occur, which will result in a mass diffusion. If we simply recall the process, the diffusion flux from the cell A to cell B within a numerical grid requires the multiplication of the area ($a$) times concentration gradient ($u$) times the diffusion coefficient ($c$) at the cell faces (Fig. 8.3).

The challenge is mostly in the calculation of the diffusion coefficient. Basically, it can be assumed that the diffusion term is either of molecular or turbulent nature. In the latter one, the mass transport is due to the mixing action of the turbulent velocity fluctuations, where the diffusion coefficient is calculated through the turbulent viscosity. In the molecular diffusion case, the diffusion is due to gradients of species concentration (mass diffusion), or because of temperature gradients (thermodiffusion).

Fick's Law, which is also called the dilute approximation, is usually applied to solve this term, where each species has either the same or a different mass diffusion constant. For more detailed diffusion processes, the multicomponent approach based on Maxwell's equations can be used. In this approach, a separate binary diffusion coefficient for each of the species combinations is considered. The diffusion component depends on the gradient of the computed species field at the inlet; therefore it needs to be defined as input data using mass or mole fractions of the species. The concentrations of the species, temperatures as well as the distribution of the fluid are often the state variables of interest in this kind of analyses.

In the following example, a typical thermochemically reacting species transport problem has been demonstrated.

### 8.2.1 3D Multiphysics Simulation of a Reformer Component

A reformer component has been used to convert methane into $H_2$ and CO by reforming part of the steam-methane fuel gas mixture. The fuel gas considering specified temperature and gas composition is released into the reformer. Air is used to heat the reformer component. Using multiphysics simulation, the fluid flow, heat transport and chemically reacting species transport within the reformer shown in Fig. 8.4 are determined.

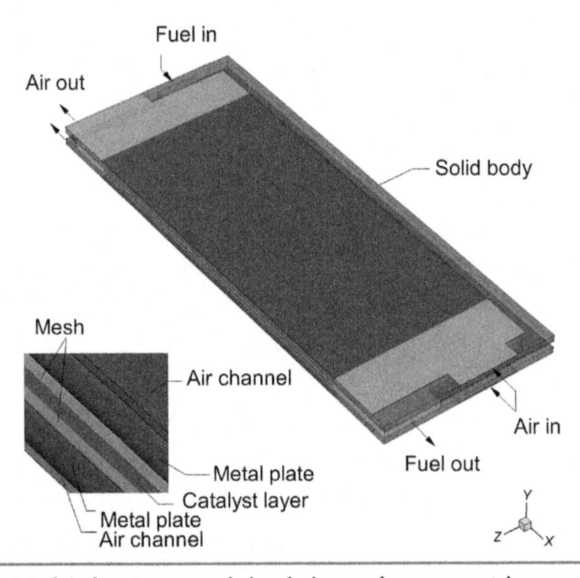


**Figure 8.4** Multiphysics model of the reformer with applied boundary conditions.

### 8.2.2 Assumptions and Approach

The reaction scheme for the present study assumes to comprise the methane-steam reforming reaction, expressed as:

$$CH_4 + H_2O \rightleftarrows CO + 3H_2 \Delta_R H^0_{298K} = 206\ \mathrm{kJ/mol} \tag{8.1}$$

and the water–gas shift reaction that is defined as:

$$CO + H_2O \rightleftarrows CO_2 + H_2 \Delta_R H^0_{298K} = -41.2\ \mathrm{kJ/mol} \tag{8.2}$$

It should be noted that the reforming reaction is endothermic and the overall reaction balance needs heat. This is provided in this example by the introduced air flow. To give the reader a bricf overview about the mathematical foundation, some details are given. Further details can be obtained from references such as [18–21]. The reaction is assumed to be at the catalyst layer. As the catalyst has been defined as a pseudofluid (porous media), the chemical system has been assumed as homogenous. The chemical reactions in this example are handled using a typical differential rate law, relating the rate of reaction to the rate of change in concentrations of the species. The kinetics and reactions are in practice chosen based on experimental data, thus it is important to derive the behaviour using these. The creation and destruction of chemical species are treated as source term. Generally, the chemical reaction kinetics can be expressed as:

$$r_i = \boldsymbol{k}_{\mathbf{f}} \prod_{j=1}^{\mathbf{N_{sp}}} |\mathbf{C}_j|^{\eta'} - k_b \prod_{j=1}^{\mathbf{N_{sp}}} |\mathbf{C}_j|^{\eta''} \tag{8.3}$$

where $\mathbf{N_{sp}}$ is the number of the chemical species in the reaction, $\mathbf{C}_j$ the molar concentration of each reactant and product, $\eta'$ is the forward rate exponent of each reactant and product species and $\eta''$ is the backward rate exponent. According the general description, the reforming rate can then be described as:

$$r = -k_f[CH_4][H_2O] - k_b[CO][H_2]^3 \tag{8.4}$$

$$r = -[CH_4][H_2O] - \frac{k_b}{k_f}[CO][H_2]^3 \tag{8.5}$$

The equilibrium constants $k_f$(forward) and $k_b$(backward) can be formulated as a concentration equilibrium constant $kx$, so that the reaction kinetics are only functions of the forward reaction. In this way, the reaction can be expressed as:

$$r = -[CH_4][H_2O] - \frac{1}{k_x}[CO][H_2]^3 \tag{8.6}$$

where $k_x$ can be abstracted as:

$$k_x = e\left(-\frac{hfi + T\Delta S_0}{RT}\right)\left(\frac{P}{RT}\right)^{\sum_{i=1}^{N}(vi,r''-vi,r')} \tag{8.7}$$

This can be used to put the whole description within an Arrhenius form of equation by using an equilibrium constant that can be described in terms of partial pressures of the species such as:

$$-r = k_r\left(P_{CH_4}P_{H_2O} - \frac{P_{CO}P_{H_2}^3}{K_P P_0^2}\right)e^{\left(\frac{-Ea}{RT}\right)} \tag{8.8}$$

The equation of state for ideal gases can be used to express all the pressure values in terms of concentration. The employed multiphysics model for this example comprises typical components of a reformer such as air channels, mesh structure, solid frame and the catalyst layer. Apart of the solid frame, each component of the system is treated as a porous zone, which is modelled as a continuum phase.

### 8.2.3 Applied Boundary Conditions and Simulation Results

First of all, the species involved in the fuel mixture needs to be specified. The properties of all species have to be introduced. If $X$ species are present, $X-1$ equations will be solved. The concentration of the $X$th species follows from the fact that all mass fractions should sum to unity. The fuel in diffusion flux depends on the concentration gradient, thus the value cannot be predicted beforehand. Therefore the modelled fluid inlet regions are specified as mass

**TABLE 8.1** Used Input Data

| | Air In | Fuel In |
|---|---|---|
| Mass flow rate (kg/h) | 30 | 2.55 |
| Temperature (°C) | 723 | 155 |
| $CH_4$ (kg/kg) | | 0.21 |
| $H_2O$ (kg/kg) | | 0.58 |
| $CO_2$ (kg/kg) | | 0.1446 |
| CO (kg/kg) | | 0.0118 |
| $H_2$ (kg/kg) | | 0.032 |

flow rates and operation temperatures together with species concentrations. Outlet can be specified as mass fraction, in case backflow occurs. Air is used for the second inlet region specified as air in. The fluid flow has been assumed to be as laminar flow. The analysis is performed in steady state for the solution of fluid flow, heat transfer, species transport and chemical reactions. Table 8.1 illustrates the used fuel in and air in boundary conditions.

Fig. 8.5 demonstrates some sample results of the analysis that gives the reader an idea about the postprocessing activities performed to analyse and understand thermochemical simulations. As an analyst it is important to be able to make interpretations from the acquired data.

For example, the thermal field results reveal that there appear thermal gradients within the catalyst layer. As the temperature distribution affects the chemical reactions within the catalyst layer, it is of paramount importance to understand the $CH_4$ distribution. The mass fraction of the methane at the fuel out region is lower compared to the inlet vicinity. This is expected because due to the chemical reactions methane is converted into hydrogen. Therefore less methane is retained at the outlet region, indicating that more heat is absorbed from the prereformer. Measurement data obtained through gas chromatography tests are often used to compare the species concentrations at the fuel out

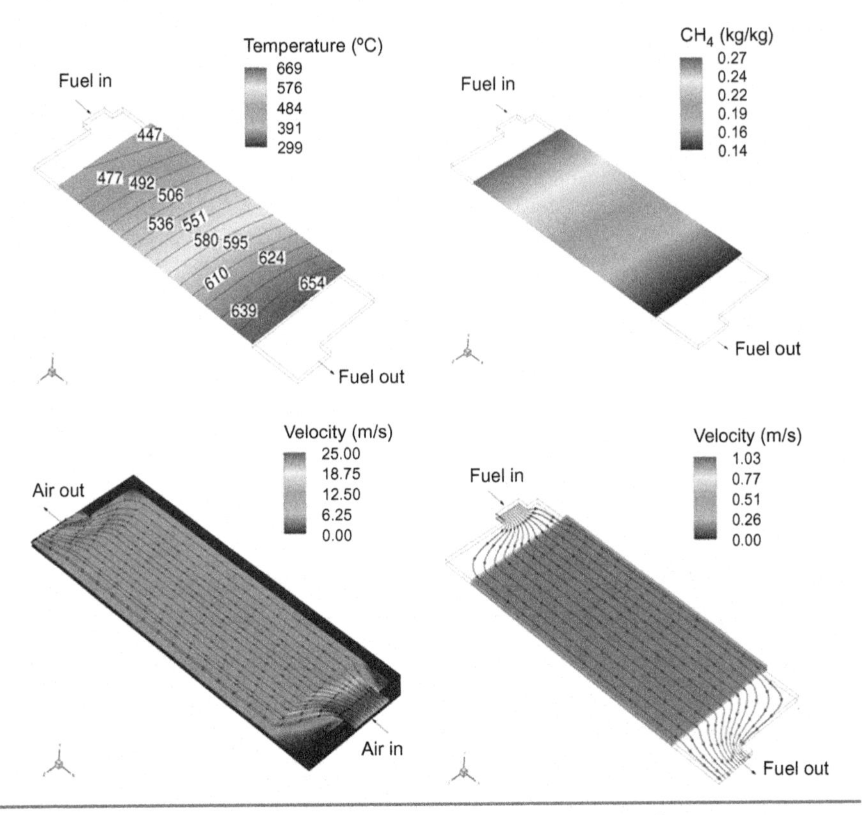


**Figure 8.5** Sample postprocessing results of the simulation.

region and gather information about the individual species of interest such as illustrated in Fig. 8.6.

## 8.3 Thermoelectrochemical Processes

Electrochemical processes are involved to a significant extent in the present-day energy economy. A major theme involves the simultaneous treatment of many complex multiphysics phenomena [22,23]. Electrochemical applications such as electroplating or electrorefining are well known, whereas the selected application fitting in this chapter will be fuel cells and batteries, as they have reached an extended use [24–32]. The goal of this chapter has been to treat energy systems from a practical point of view, thus emphasis has been placed on the

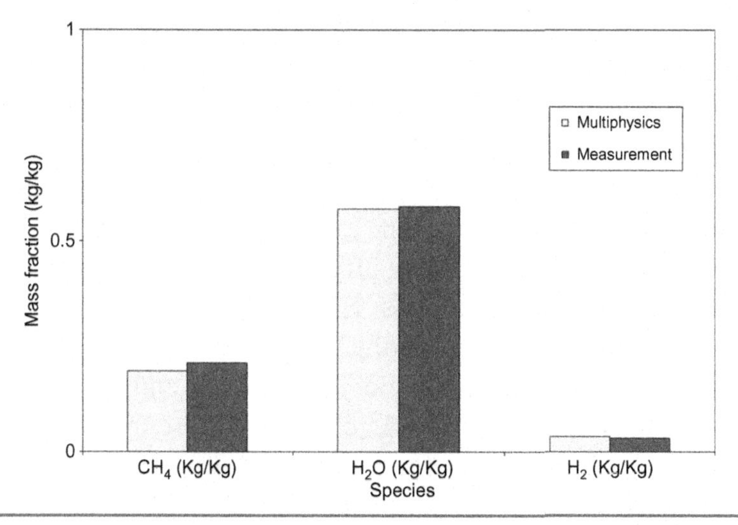


**Figure 8.6** Experimental validation of species concentrations.

modelling of these electrochemical devices. The understanding of electrochemical processes requires an immense knowledge and fundamentals, including the thermodynamics, electrode kinetics, as well as transport phenomena. In this circumstance, each application would require a complete discussion. Therefore it has been pursued to give a very brief summary of the modelling requirements and proceed by examples. This will provide an introduction to the multiphysics simulation of these devices, as they become economically within the energy and automotive sectors important.

### 8.3.1 3D Multiphysics Simulation of a Li-Ion Battery Pack

Batteries are one of the most important sources for energy storage. Particularly, in the development of Hybrid Electrical Vehicles and Electrical Vehicles they play a significant role. Among the batteries, lithium-based ones receive the most attention due to their high voltage, favourable energy density and low self-discharge rates. Although Lithium Ion batteries show great performance, they are prone to thermal effects like it has been observed during continuous charging, discharging and working under high temperatures that leads to cell degradation.

Multiphysics simulation at cell and pack level aids to contribute to a better understanding of the thermal effects and provides detailed information about other field variables that play a significant role in design purposes. As in most of the multiphysics problems, the multi-dimensional approach is also in the simulation of cell and pack level battery simulations important. The potential drop along the current collector due to Ohmic drop may affect the current distribution, with a higher current closer to the tabs. Hence the battery management is supported in terms of the state of the battery life.

Therefore the current example demonstrates a multiphysics simulation of a three cell battery pack, connected in series (1P3S as we would call it, where P stands for parallel and S for series). The electrodes are assumed to be made of the same material for simplicity. The electrical parameters are chosen such that a discharging process of the battery pack is assumed to occur under a constant power of 300 W. The nominal cell capacity is assumed to be 20 Ah. The electric conductivities of the positive and negative electrodes are assumed to be the same (same material) and a value of $3.0E^{+07}$ 1/ohm-m has been chosen. Some analyst may prefer do define diffusion coefficients and use them instead of electric conductivities.

Natural convection has been assumed on the walls with a heat transfer coefficient of 8 W/m$^2$-K and a temperature of 293K. The three-dimensional temperature distribution and current density as well as the total heat generation has been calculated within the LIB module as a function of a demonstration time of 200 s.

To solve the problem, the thermal and electric fields are solved using the following equations:

$$\frac{\partial \rho C_p T}{\partial t} - \nabla .(k \nabla T) = \kappa_+ \left| \nabla \phi_+ \right|^2 + \kappa_- \left| \nabla \phi_- \right|^2 + \dot{q}_{ec} + \dot{q}_s + \dot{q}_a \quad (8.9)$$

$$\begin{aligned} \nabla .(\kappa_+ \nabla \phi_+) &= -(j_{ec} - j_s) \\ \nabla .(\kappa_- \nabla \phi_-) &= j_{ec} - j_s \end{aligned} \quad (8.10)$$

where $\kappa_-$ and $\kappa_+$ refer to the effective electric conductivities of the electrodes; $\phi_-$ and $\phi_+$ are the phase potentials of the electrodes. The source term $\dot{q}_{ec}$, representing the heat generation due to

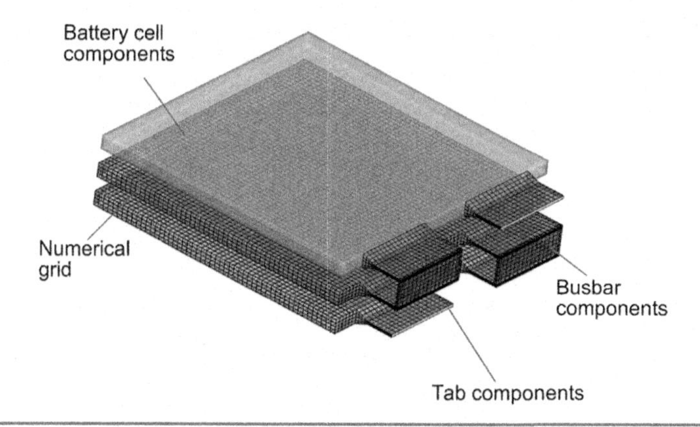


**Figure 8.7** Multiphysics model of the battery: components and part of the numerical grid.

electrochemical reactions and the volumetric current transfer rate $j_{ec}$ are calculated using an electrochemical model [33,34]. The electrochemical models can be substituted using any experimentally determined model as it is a source term. $j_s$ and $\dot{q}_{ec}$ are the current transfer rate and heat generation due to the battery internal short-circuit, respectively. Of course these two terms can be ignored if no short-circuit is present. The same implies for the abuse term $\dot{q}_a$.

The used computational model has been illustrated in Fig. 8.7.

In this model the battery is assumed to be a homogeneous domain (does not require resolved battery components) and it is important to create the zones active domain cell and the passive domains named as tab and busbar. These domains need to be carefully defined as the electrochemical reactions will only be present on the active zones, whereas the electric potential field is solved in both regions. The used thermal properties given in Table 8.2 are from the literature [35].

Fig. 8.8 illustrates the predicted thermal results at a discharge time of 200 s. The temperature distribution is close to 300 K with some differences of 2–3 K around the tab and busbar regions. The reader should consider that in a practical case the convective boundary conditions used in this simulation may not reflect the convective heat transfer conditions to which the walls of the module is exposed

**TABLE 8.2** Used Thermal Properties

| Thermal Properties | Battery |
|---|---|
| $\rho$ (kg/m$^3$) | 2765 |
| $Cp$ (J/kg-K) | 1394 |
| $k_x$ (W/m-K) | 25.7 |
| $k_y$ (W/m-K) | 25.7 |
| $k_z$ (W/m-K) | 0.794 |

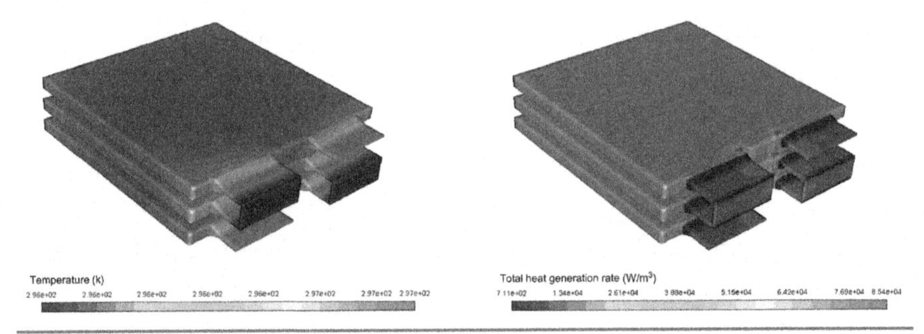


**Figure 8.8** Thermal behaviour of the battery: temperature and total heat generation rates.

to. In the current model, the surrounding walls of the model are exposed to uniform convection and an active fluid flow has not been considered. The temperature distribution may also vary during or close to the end of the whole discharge process. This may be the case when actively forced convection would be the cooling method and the volumetric flow rate of cooling air would be decreased or increased. This may affect the temperatures of multiple cells located inside a module. The total heat generation contour plot exemplifies the predicted results that show in detail the region between the tab components and the cell as generating the highest amount of heat.

Likewise, Fig. 8.9 shows the charge flow in amperes per unit area of cross section. This is a vector quantity having both a magnitude (scalar) and a direction, thus the vector field plot has been illustrated.

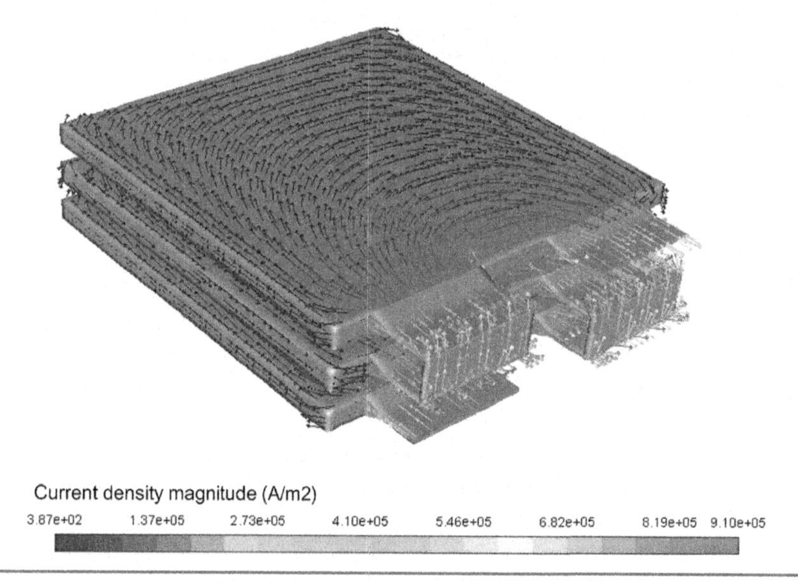


**Figure 8.9** Vector plot of the current density inside the battery pack.

The analyst can perform simple verification procedures like in mechanics, problems using simple models. Equations describing the current density such as $J = I/A$ can be used to become a feeling of how many amperes of current is flowing through the battery in a given area. Likewise, it is possible to perform simple verifications using field variables. The total heat generation rate is the sum of the volumetric Ohmic source, the electrochemical source and the short circuit source. When considering the current example, the approximate values given in Table 8.3 would be obtained.

## 8.4 Thermomechanical Processes

The subject thermomechanical modelling has been handled in detail in previous chapters. Most of the components and systems in energy technologies are subjected to high temperatures. Therefore the thermomechanical behaviour of the used materials, components, as well as their interactions affects significantly the overall system behaviour. The following example should aid to emphasise the importance of employing thermomechanical analyses in energy systems as well.

**TABLE 8.3** Tabulated Example Results

| Field Variable | |
|---|---|
| Volumetric Ohmic source | 0.429 W |
| Electrochemistry source | 5.015 W |
| Short circuit source | 0 W (very high resistance was chosen-expected) |
| Total heat generation | 5.44 W |
| Potential Phi + battery tab voltage | 11.73 V |
| Battery tab current | 63.9 A |

For this purpose, a thermomechanical analysis of a full solid oxide fuel cell (SOFC) component has been demonstrated.

SOFC systems are used in stationary and auxiliary power units in various engineering fields, including marine, aerospace as well as automotive engineering. The challenge of modelling SOFC components is the highly nonlinear geometrical features, ranging from micrometre to metre scale, thus a multiscale problem is present. Moreover, the use of various materials [36,37] with different function and temperature dependency requires also to account for nonlinear material effects. Highly fragile glass-ceramic components used as sealants, fuel cell components such as ceramic composite electrodes, as well as metal components are examples [37–40]. As SOFCs operate at temperatures around 750°C, the complex material behaviour is important. One of the most challenging parts in modelling SOFC components and systems has been the inclusion of the complex processes of the system components [41–43]. Governing equations for fluid flow, heat transfer, electric field as well as chemically reacting species transport and source terms of electrochemistry are usually solved to provide the thermal field that is used for complex thermomechanical analyses [44–53].

### 8.4.1 3D Multiphysics Simulation of a Solid Oxide Fuel Cell Stack

The following example should demonstrate a thermomechanical analysis of a whole SOFC component, including 36 layers that is known as a 'stack'. To focus on the thermomechanical subject as well as to sparingly use the allocated space for this section, details of the complex thermofluid analysis has not been elucidated, but the utilised thermal field will be illustrated (Fig. 8.10). For more details about the thermofluid flow process inside the stack model, the reader may refer to Peksen [54,55].

Fig. 8.11 illustrates the details of the used multiphysics model. The stack encompasses 36 fuel cell layers, each comprising the cell units, sealant components and metal interconnector and frame units. The full component shows symmetrical behaviour, thus half of the stack can be utilised for the analysis. The assumption to consider half of the body needs to be meaningful. Therefore appropriate boundary conditions need to be applied, in order that the solver can comprehend that another half of the component is actually present. This is achieved by applying symmetry boundary conditions to the region where the model has been divided.

Fig. 8.12 exemplifies the thermomechanically induced stress results of the interconnector metal and sealant components. Mind

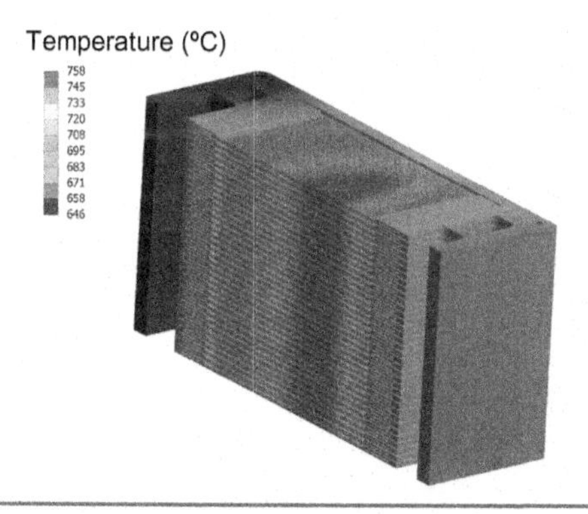


**Figure 8.10** Used thermal field result as thermal boundary condition.

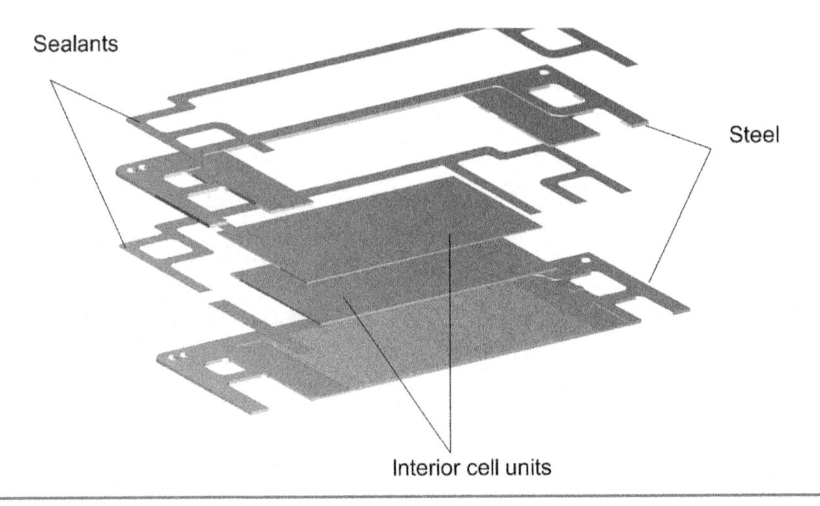


**Figure 8.11** Employed model details: subcomponents and model assumptions.

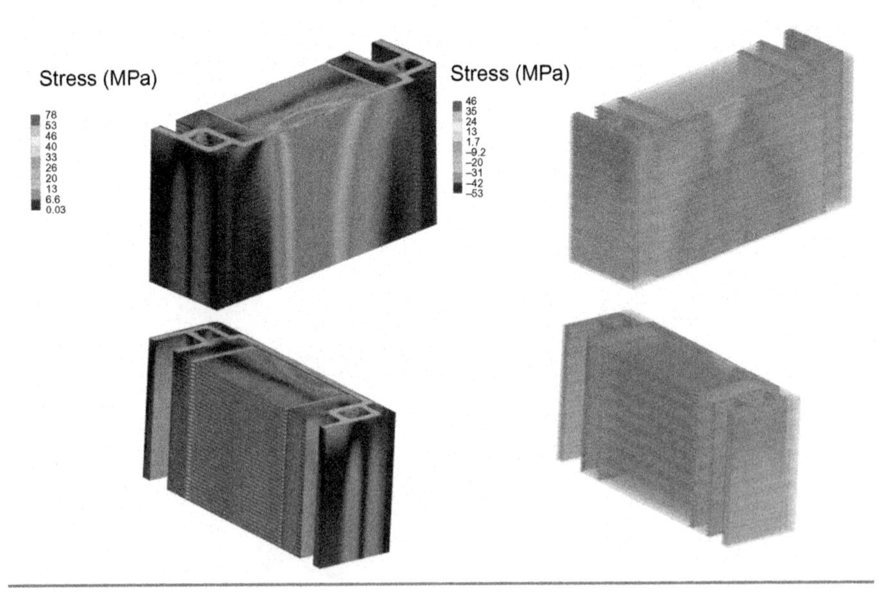


**Figure 8.12** Stress field distribution of the stack: metal interconnector plates (left); sealant components (right).

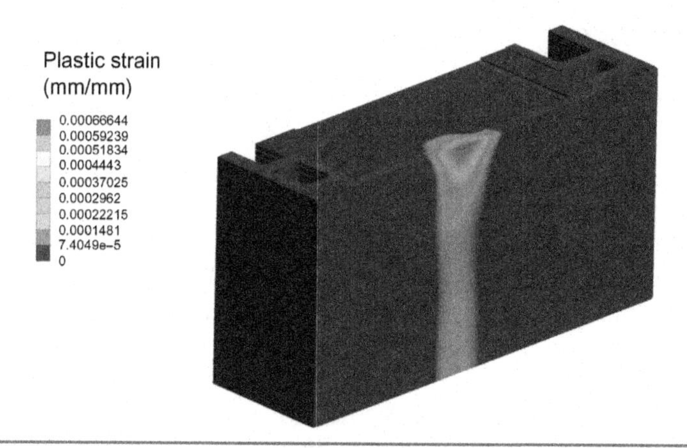


**Figure 8.13** Plastic strain results of the interconnector plates.

that the metal components are interpreted using the Von Mises stress theory suitable for ductile materials. This is the reason why all values in the legend are positive. The stress distribution for the sealants has been calculated based on the maximum principal stress theory, explaining the presence of both negative and positive signs.

The results reveal that the stack shows the highest stress values at the mid region facing outward (and inside the manifold ports that are not illustrated in detail). The reason is that the stack is only able to elongate outward and upward because of the constraint and mechanical loading. As the components are joined together and the elongation is limited, stresses arise. The stress field pattern illustrated on the top region of the sealant materials is of great importance. The analyst must be able to explain this kind of behaviour. To explain this, Fig. 8.13 illustrates the plastic strain results of the metal components.

The plastic strain results show that the metal has reached its yield point and started to continuously deform. This is most prominent on the top region of the component. In practice this means that the metal components elongate at this region the most. However, the sealant materials are rigidly bonded with the interconnector plates. Therefore the deformation of the metal components is restricted by the sealant materials. As soon as this limit has been achieved, the effect of the deformation in the metal results in stress in the sealants, explaining the stress pattern matching this strain profile.

## References

[1] Sirignano WA. Fluid dynamics and transport of droplets and sprays. Cambridge: Cambridge University Press; 2010.

[2] Numerical modeling of miniature cyclone. Powder Technol 2017;320:325–39. Available from: https://doi.org/10.1016/J.POWTEC.2017.07.053.

[3] Analysis of single phase, discrete and mixture models, in predicting nanofluid transport. Int J Heat Mass Transf 2017;114:225–37. Available from: https://doi.org/10.1016/J.IJHEATMASSTRANSFER.2017.06.030.

[4] DPM simulation in an underground entry: comparison between particle and species models. Int J Min Sci Technol 2016;26:487–94. Available from: https://doi.org/10.1016/J.IJMST.2016.02.018.

[5] Anantpinijwatna A, Sin G, O'Connell JP, Gani R. A framework for the modelling of biphasic reacting systems. Comput Aided Chem Eng 2014;34:249–54. Available from: https://doi.org/10.1016/B978-0-444-63433-7.50026-2.

[6] Andrianopoulos E, Korre A, Durucan S, Franzsen S. Coupled Thermo-Mechanical-Chemical modelling of underground coal gasification. Comput Aided Chem Eng 2016;38:1069–74. Available from: https://doi.org/10.1016/B978-0-444-63428-3.50183-1.

[7] Bahadori A, Bahadori A. Chapter 3 – Single-phase and multiphase flow in natural gas production systems. In: Nat. Gas Process. 2014; 59–150. doi:10.1016/B978-0-08-099971-5.00003-9.

[8] Boyadjiev CB, Babak VN, Boyadjiev CB, Babak VN. PART 3 – Chemically reacting gas-liquid systems. Non-Linear Mass Transf. Hydrodyn. Stab. 2000;171–223. Available from: https://doi.org/10.1016/B978-044450428-9/50004-1.

[9] Carter JG, Cokljat D, Blake RJ, Westwood MJ. Computation of chemically reacting flow on parallel systems. Parallel Comput Fluid Dyn 1995;1996:113–20. Available from: https://doi.org/10.1016/B978-044482322-9/50068-2.

[10] Chattopadhyay K, Guthrie RIL. Chapter 4.6 – Single phase, two phase, and multiphase flows, and methods to model these flows. In: Treatise Process Metall. 2014; 527–53. Available from: https://doi.org/10.1016/B978-0-08-096984-8.00012-4.

[11] Chin WC, Chin WC. Chapter 11 – Effective properties in single and multiphase flows. In: Quant. Methods Reserv. Eng. 2017; 235–45. Available from: https://doi.org/10.1016/B978-0-12-810518-4.00011-6.

[12] Chin WC, Chin WC. Chapter 21 – Forward and inverse multiphase flow modeling. In: Quant. Methods Reserv. Eng. 2017; 587–639. Available from: https://doi.org/10.1016/B978-0-12-810518-4.00021-9.

[13] De Bortoli ÁL, Andreis GSL, Pereira FN, De Bortoli ÁL, Andreis GSL, Pereira FN. Chapter 4 – Mixing and turbulent flows. In: Model. Simul. React. Flows 2015; 53–72. Available from: https://doi.org/10.1016/B978-0-12-802974-9.00004-0.
[14] Hernández A, Hernández A. Chapter 3 – Multiphase flow. In: Fundam. Gas Lift Eng. 2016; 81–126. Available from: https://doi.org/10.1016/B978-0-12-804133-8.00003-8.
[15] Huinink H. Two phase flow. Fluids in porous media. Morgan & Claypool Publishers; 2016. p. 9–15. Available from: https://doi.org/10.1088/978-1-6817-4297-7ch9.
[16] Sieniutycz S, Sieniutycz S. Chapter 12 – Multiphase flow systems. In: Thermodyn. Approaches Eng. Syst. 2016; 561–82. Available from: https://doi.org/10.1016/B978-0-12-805462-8.00012-1.
[17] Chapter 3 – Modern notions of two-phase flows. In: Entropic invariants of two-phase flows; 2015. p. 37–58. Available from: https://doi.org/10.1016/B978-0-12-801458-5.00003-4.
[18] Nagy E. Basic equations of the mass transport through a membrane layer. Elsevier; 2012.
[19] Numerical modelling and experimental validation of a planar type pre-reformer in SOFC technology. Int J Hydrogen Energy 2009;34:6425–36. Available from: https://doi.org/10.1016/J.IJHYDENE.2009.06.017.
[20] Optimisation of a solid oxide fuel cell reformer using surrogate modelling, design of experiments and computational fluid dynamics. Int J Hydrogen Energy 2012;37:12540–7. Available from: https://doi.org/10.1016/J.IJHYDENE.2012.05.137.
[21] Investigation of methane steam reforming in planar porous support of solid oxide fuel cell. Appl Therm Eng 2009;29:1106–13. Available from: https://doi.org/10.1016/J.APPLTHERMALENG.2008.05.027.
[22] Electrochemistry; n.d.
[23] Hamnett A. Fundamentals of electrochemistry; 2001. p. 270–97. Available from: https://doi.org/10.3969/j.issn.2095-4239.2013.03.011.
[24] Bucci G, Swamy T, Bishop S, Sheldon BW, Chiang Y-M, Carter WC. The effect of stress on battery-electrode capacity. J Electrochem Soc 2017;164: A645–54. Available from: https://doi.org/10.1149/2.0371704jes.
[25] Carmo M, Fritz DL, Mergel J, Stolten D. A comprehensive review on PEM water electrolysis. Int J Hydrogen Energy 2013;38:4901–34. Available from: https://doi.org/10.1016/j.ijhydene.2013.01.151.
[26] Gandía LM, Arzamendi G, Diéguez PM, Martín AJ, Hornés A, et al. Chapter 15 – Recent advances in fuel cells for transport and stationary applications. In: Renew. Hydrog. Technol. 2013; 361–380. Available from: https://doi.org/10.1016/B978-0-444-56352-1.00015-5.

[27] Manuel Stephan A, Nahm KS. Review on composite polymer electrolytes for lithium batteries. Polymer (Guildf) 2006;47:5952–64. Available from: https://doi.org/10.1016/j.polymer.2006.05.069.

[28] Mertens A, Vinke IC, Tempel H, Kungl H, de Haart LGJ, et al. Quantitative analysis of time-domain supported electrochemical impedance spectroscopy data of Li-ion batteries: reliable activation energy determination at low frequencies. J Electrochem Soc. 2016;163:H521–7. Available from: https://doi.org/10.1149/2.0511607jes.

[29] Spiegel C, Spiegel C. Fuel cell electrochemistry. P.E.M. Fuel Cell Model. Simul. Using Matlab. Elsevier; 2008. p. 49–76. Available from: https://doi.org/10.1016/B978-012374259-9.50004-5.

[30] Xia Y, Wierzbicki T, Sahraei E, Zhang X. Damage of cells and battery packs due to ground impact. J Power Sources 2014;267:78–97. Available from: https://doi.org/10.1016/j.jpowsour.2014.05.078.

[31] Xu J, Liu B, Wang L, Shang S. Dynamic mechanical integrity of cylindrical lithium-ion battery cell upon crushing. Eng Fail Anal. 2015;53:97–110. Available from: https://doi.org/10.1016/j.engfailanal.2015.03.025.

[32] Yilanci A, Dincer I, Ozturk HK. A review on solar-hydrogen/fuel cell hybrid energy systems for stationary applications. Prog Energy Combust Sci 2009;35:231–44. Available from: https://doi.org/10.1016/j.pecs.2008.07.004.

[33] Kwon KH, Shin CB, Kang TH, Kim CS. A two-dimensional modeling of a lithium-polymer battery. J Power Sources 2006. Available from: https://doi.org/10.1016/j.jpowsour.2006.03.012.

[34] Kim G-H, Smith K, Lee K-J, Santhanagopalan S, Pesaran A. Multi-domain modeling of lithium-ion batteries encompassing multi-physics in varied length scales. J Electrochem Soc 2011;158:A955–69. Available from: https://doi.org/10.1149/1.3597614.

[35] Yi J, Koo B, Shin CB. Three-dimensional modeling of the thermal behavior of a lithium-ion battery module for hybrid electric vehicle applications. Energies 2014;7:7586–601. Available from: https://doi.org/10.3390/en7117586.

[36] Nguyen XV, Chang CT, Bin JG, Chan SH, Lee WT, Chang SW, et al. Study of sealants for SOFC. Int J Hydrogen Energy 2016;41:21812–19. Available from: https://doi.org/10.1016/j.ijhydene.2016.07.156.

[37] Mahato N, Banerjee A, Gupta A, Omar S, Balani K. Progress in material selection for solid oxide fuel cell technology: a review. Prog Mater Sci 2015;72:141–337. Available from: https://doi.org/10.1016/j.pmatsci.2015.01.001.

[38] Liu M, Lynch ME, Blinn K, Alamgir FM, Choi Y. Rational SOFC material design: new advances and tools. Mater Today 2011;14:534–46. Available from: https://doi.org/10.1016/S1369-7021(11)70279-6.

[39] Sartori F, Silva D, Ofilo T, De Souza M. Novel materials for solid oxide fuel cell technologies: a literature review. Int J Hydrogen Energy 2017;42:26020–36. Available from: https://doi.org/10.1016/j.ijhydene.2017.08.105.

[40] Ivers-Tiffée E, Weber A, Herbstritt D. Materials and technologies for SOFC-components. J Eur Ceram Soc 2001;21:1805–11. Available from: https://doi.org/10.1016/S0955-2219(01)00120-0.

[41] Fang Q, Blum L, Peters R, Peksen M, Batfalsky P, Stolten D. SOFC stack performance under high fuel utilization. Int J Hydrogen Energy 2015;40:1128–36. Available from: https://doi.org/10.1016/j.ijhydene.2014.11.094.

[42] Blum L, Groß SM, Malzbender J, Pabst U, Peksen M, Peters R, et al. Investigation of solid oxide fuel cell sealing behavior under stack relevant conditions at Forschungszentrum Jülich. J Power Sources 2011;196:7175–81. Available from: https://doi.org/10.1016/j.jpowsour.2010.09.041.

[43] Peters R, Deja R, Engelbracht M, Frank M, Nguyen VN, Blum L, et al. Efficiency analysis of a hydrogen-fueled solid oxide fuel cell system with anode off-gas recirculation. J Power Sources 2016;328:105–13. Available from: https://doi.org/10.1016/j.jpowsour.2016.08.002.

[44] Peksen M, Peters R, Blum L, Stolten D. 3D coupled CFD/FEM modelling and experimental validation of a planar type air pre-heater used in SOFC technology. Int J Hydrogen Energy 2011;36:6851–61.

[45] Peksen M, Meric D, Al-Masri A, Stolten D. A 3D multiphysics model and its experimental validation for predicting the mixing and combustion characteristics of an afterburner. Int J Hydrogen Energy 2015;40:9462–72. Available from: https://doi.org/10.1016/j.ijhydene.2015.05.103.

[46] Peksen M. 3D thermomechanical behaviour of solid oxide fuel cells operating in different environments. Int J Hydrogen Energy 2013;38 13408–18.

[47] Peksen M, Peters R, Blum L, Stolten D. Numerical modelling and experimental validation of a planar type pre-reformer in SOFC technology. Int J Hydrogen Energy 2009;34:6425–36.

[48] Al-Masri A, Peksen M, Blum L, Stolten D. A 3D CFD model for predicting the temperature distribution in a full scale APU SOFC short stack under transient operating conditions. Appl Energy 2014;135:539–47. Available from: https://doi.org/10.1016/j.apenergy.2014.08.052.

[49] Andersson M, Yuan J, Sundén B. SOFC modeling considering electrochemical reactions at the active three phase boundaries. Int J Heat Mass Transf 2012;55:773–88. Available from: https://doi.org/10.1016/j.ijheatmasstransfer.2011.10.032.

[50] Amiri A, Vijay P, Tadé MO, Ahmed K, Ingram GD, Pareek V, et al. Planar SOFC system modelling and simulation including a 3D stack module. Int J Hydrogen Energy 2016;41:2919–30. Available from: https://doi.org/10.1016/j.ijhydene.2015.12.076.

[51] Kakaç S, Pramuanjaroenkij A, Zhou XY. A review of numerical modeling of solid oxide fuel cells. Int J Hydrogen Energy 2007;32:761–86. Available from: https://doi.org/10.1016/j.ijhydene.2006.11.028.

[52] Kendall K, Kendall M, Pianko-Oprych P, Jaworski Z, Kendall K. High-temperature solid oxide fuel cells for the 21st century. Elsevier; 2016. Available from: https://doi.org/10.1016/B978-0-12-410453-2.00013-0.

[53] Khaleel MA, Selman JR. Chapter 11 – Cell, Stack and System Modelling. In: High Temp Solid Oxide Fuel Cells; 2003. p. 291–331. Available from: https://doi.org/10.1016/B978-185617387-2/50028-3.

[54] Peksen M. A coupled 3D thermofluid-thermomechanical analysis of a planar type production scale SOFC stack. Int J Hydrogen Energy 2011;36:11914–28.

[55] Peksen M. Numerical thermomechanical modelling of solid oxide fuel cells. Prog Energy Combust Sci 2015;48:1–20. Available from: https://doi.org/10.1016/j.pecs.2014.12.001.

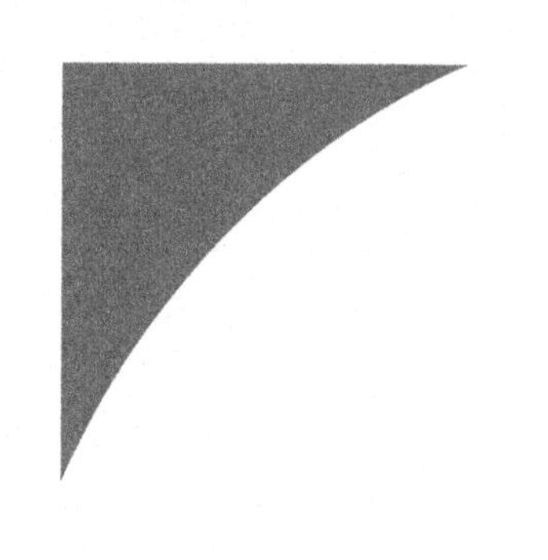

Chapter 9

# Multiphysics Modelling Issues

## Chapter Outline



Multiphysics modelling has become an essential and powerful tool in most branches of science and engineering. To confirm its effectiveness, multiphysics modelling utilised at least three major phases throughout this book. First of all, the problem has been physically and mathematically described. Second, the analyst has devised advanced techniques for solving the complex governing equations, thus numerical methods have been employed. Ultimately, the first



Multiphysics Modelling. DOI: https://doi.org/10.1016/B978-0-12-811824-5.00009-2

two phases would not remain separately, thus the multiphysics analyst must marshal a thorough understanding of the interactions between the multiprocesses of the system, as well as their numerical analogy to ensure that the quantitative predictions reflect the modelled reality. The importance of validation–verification has been emphasised for this purpose.

Traditionally, these fields have been separated into distinct disciplines, including various types of processes (the term process is used on purpose, as not to bound the understanding to branches solely of continuum mechanics), numerical analysis, etc. Indeed each field is very deep in its own right; however, their combination of viewpoints results in a powerful approach, i.e. multiphysics to comprehend and predict natural phenomena.

It is this combination the book efforts to meld all aspects of each task into a unified treatment. Each chapter has been treated sparingly to serve the purpose, which is to provide a consistent framework for an understanding and approach of modelling multiphysics phenomenon. However, there are many challenges and issues, whilst multiphysics modelling. In this chapter, it has been targeted to give a brief overview of these challenges and issues most analysts face. These challenges will influence the accuracy of the simulation results and can be due to sources of different nature. The multiphysics analyst needs to improve the understanding of the cause of each different issue type, as to develop strategies and best practices to minimise the output errors.

In the previous chapters, the effects of geometrical model consideration and assumptions, loading, constraints, as well as numerical grid have been exemplified in detail. The used computational domain for the solution of the complex problem has been shown to influence the accuracy and reliability of the predicted results. The effects and limits of the assumptions and simplifications on the solution of the analysis has been another important physical aspect.

When these kind of error sources interact with software–hardware or algorithm problems, the desired output obtained from a multiphysics modelling approach will be on a poor side. To mitigate these problems several aspects need to be considered. In addition to the previously mentioned problems, the current chapter aims to shed light on some important numerical modelling issues often seen in practice.

There are several different factors that combine to affect the overall solution accuracy throughout the multiphysics analysis. In addition to the previously mentioned issues, five of them are predominantly seen, thus will be considered, as it is not possible to cover the entire range of issues observed in multiphysics solutions. The reader may refer to subject-specific journals such as *Journal of Computational Physics*, *Acta Mathematica Scientia*, *Journal of Systems Architecture*, *Journal of Parallel and Distributed Computing*, *Advances in Engineering Software*, *Computer Standards & Interfaces* and *Procedia Computer Science* or for understanding specific topics such as large data storage textbooks such as Linstedt [1]. The current issues introduced in this chapter that lead to multiphysics modelling issues are categorised as follows:

## 9.1 Performance Limits

Several computational and engineering requirements have been listed over the recent years to express the strong demand on high-performance computing, which led to the term parallel computing or known as supercomputing. These have been evolved thanks to the developments in the microelectronics technology. Assuming the use of the fastest technology, supercomputers typically use parallelism in the form of either vector processing or array processing to obtain performance. Supercomputers and their developments have been a topic since the 1960s so that they have been regularly ranked [2–4]. Over the years, the supercomputer technologies have progressively developed from peak performances of megaflops or gigaflops range (FLOPS is a measure of computer performance) with few hundreds of processors to petal scale and tens of thousands of commodity CPUs.

The early parallel computers such as ILLIAC IV with more than 200 processors or Cray-1 with its 160 mega floating point operations per second (Mflops) but nearly 6 t of weight have been replaced by modern machines such as the MDGRAPE-3 [5] with its ultra-high performance petascale supercomputer system level, which will probably be substituted with other machines in the near future, as well. Today, there are various vendors providing single or distributed supercomputing solutions. Exascale performance is expected to be used widely between early 2021 and late 2022 as either hybrid systems or multicore systems [6]. To reach zetta scale computing, the scientific community has still work to go; however, within the next decade this might be possible, despite the high costs of this kind of machines.

Fig. 9.1 depicts an example supercomputer, illustrating JUQUEEN-Jülich Blue Gene/Q of the Forschungszentrum Jülich, Germany. The author utilised many years former generations JUGENE and JUROPA of the 28 racks system that has been under the top five supercomputers in the world.

**Figure 9.1** Supercomputer of the new generation: with its 458,752 cores, the new JUQUEEN reaches a peak performance of 5.9 Petaflop/s—the first of its kind in Europe. *Photograph courtesy of Forschungszentrum Jülich, Germany.*

The increased demand on advanced and complex multiphysics problems has brought to the fore the need for more sophisticated hardware and compilers to parallelise existing multiphysics applications and methods on software and to store large simulation files [7,8]. The complexity of the problem increases in size and resolution; therefore it is necessary to utilise a parallel dynamic code on parallel systems [9,10]. These in conjunction with new and faster algorithm libraries led to developments in various computational fields. Moreover, due to the cost advantage for building a parallel system out of commodity parts-CPUs, memory, network, etc., building a self-made parallel computer has been trendy.

The demand and choice of supercomputers depends on several factors such as hardware, system management and its distributed supercomputing capacity. Due to their high consumption of energy and release of heat, improved energy efficiencies have been an important factor in recent years [11]. Therefore the ultimate message can be attributed to the multiphysics basics, i.e. about what the analyst wants to achieve. Accordingly, the efficient solution with a suitable architecture needs to be considered.

## 9.2 Round-Off Errors

The numerically calculated values in a multiphysics analysis can be represented in a binary numeral system, which is the fundamental way of expressing numerical data in a computational platform. Since any given integer has only a finite number of digits, it is possible to express all integers below a certain limit exactly. Noninteger numbers are different, since infinitely many digits are required to represent most of them. This is regardless of the choice of the numerical system.

This is the reason why noninteger numbers cannot be expressed in a computer without committing an error. This error is referred as *round-off error* or *rounding error*. The standard computer representation of noninteger numbers is expressing as floating point numbers with a fixed number of bits. Usually, an error is not only committed when a number is represented as a floating point number. Most predictions with floating point numbers will further induce round-off errors. In most situations these errors will be small, but for high-accuracy computations round-off can be the dominating source of errors.

Several methods such as variants of the implicit Runge–Kutta or symmetric multistep methods are utilised to minimise round-off errors [12]. Several studies discuss the improvements of round-off errors for different applications [13,14]. Studies show that the greatest potential within the four elementary operations, addition, subtraction, multiplication and division with floating point numbers are observed in the addition/subtraction, which in certain situations may lead to dramatic errors. This can be exemplified as follows:

Supposedly two numbers such as $a = 5.636$ and $b = 7.830$ are considered.

If these two numbers are converted into a normal form, they would be expressed as:

$$a = 0.5636 \times 10^1 \text{ and } b = 0.7.830 \times 10^1$$

When the two numbers are added as $0.5636 + 0.7830$, the result would be 1.3466. The result in an exact arithmetic form would be expressed as $0.13466 \times 10^2$. However, this is not the normal form, since the significand (coefficient) is composed of five digits. Therefore a rounding has been performed and the value 0.13466 is rounded approximately to 0.1347. The final result would then be represented as $0.1347 \times 10^2$.

The relative error in this result would be around $3 \times 10^{-4}$. It is obvious that once normalised numbers are added and the result converted to normalised form with a fixed number of digits, errors may easily occur.

In subtraction, the error gets of significant importance as the following example will depict:

Imagine that the first value is expressed as $a = 10/7$ and the second value $b = -1.42$

The conversion in normal would yield to:

$a = 10/7 \sim 0.1429 \times 10^1$ and $b = -0.142 \times 10^1$ adding the two values would lead to $0.1429 - 0.142 = 0.0009$. When this result is converted to normal form, the results are predicted to be $0.9000 \times 10^{-3}$.

The correct value would be $0.8571 \times 10^{-3}$ when the result would be rounded to four digits, while yielding to a more significant relative error of $10^{-1}$. The starting point with two numbers with full four digit accuracy has been computed to only one correct digit. In other words, almost all the accuracy has been lost when the subtraction was performed. The computer was not able to comprehend that the additional digits to be added in $0.8571 \times 10^{-3}$ should be considered from the decimal expansion of 10/7. Hence, the accuracy loss has been reflected in the relative error.

The two samples aim to demonstrate the reader how round-off errors arise in the predictions. In practice, floating point multiplication and division do not lead to loss of correct digits as long as the numbers are within the range of the floating point model. In the worst case, the last digit in the result may be one digit wrong.

Ultimately, the multiphysics modelling issues based on round-off errors are due to the fact that only a finite amount of information can be stored at any stage of the calculation process. They are closely in relation to floating point computations. Therefore they are interacting with the limitations of the computers capacity.

## 9.3 Iteration Issues

In general, nonlinear equations in multiphysics cannot be solved in a finite sequence of steps. As linear equations can be solved using direct methods, nonlinear equations utilise iterative methods. As the reader already knows from introductory courses, in contrast to direct methods (such as Gaussian elimination) for a solution of $Au = b$, the numerical iterative methods start with an initial guess for the solution vector and produce improved and more accurate approximations via

**TABLE 9.1** Iterative Error Example

| $k$ | $x_k$ | $E_k$ |
|---|---|---|
| 0 | 1 | 0.41421356237310 |
| 1 | 1.5 | 0.08578643762690 |
| 2 | 1.41666666666667 | 0.00245310429357 |
| 3 | 1.41421568627457 | 0.00000212390141 |
| 4 | 1.41421356237469 | 0.00000000000159 |

a sequence of iterations. The iterative methods do not normally yield $u$ in a finite number of steps.

One can terminate such a calculation procedure after a finite number of sequences when it has produced a sufficiently good approximation. To that end, we consider that an iterative method generates a sequence of iterates $x_0$, $x_1$, $x_2$,... that converges to the exact solution $x_n$. The difference between individual iterates and the exact solution provided by iteration $x_n$ is the iterative error. As to remind the reader briefly, Table 9.1 exemplifies the iterative errors for the solution of:

$$f(x) = 0, \quad \text{where}$$
$$f(x) = x^2 - 2$$

The error $E_k = x_k - x^*$, where $x^* = \sqrt{2}$ is the exact solution.

The Newton's Method is used to find the solution. Examining the predicted values in the table, it is visible that the number of correct decimal places roughly doubles with each iteration, which is typical of quadratic convergence.

As it can be recognised, as the iteration number increases the iterative error decreases while getting closer to the exact solution.

In practice, mostly a variable of interest has been chosen and a residual normal be defined. Monitoring this variable against its residual curve enables to follow the behaviour.

## 9.4 Solution Errors

Solution errors are based on the accuracy of the discretisation and not on the iterative speed or residual accuracy like the iterative errors are concerned of. Thus the solution errors are dealing with the correctness of the solved problem. The used numerical code may be valid to be first- or second-order accurate, but by solving different types of problems, different level of result accuracies may occur. Thus the level of the numerical solution errors may vary.

The grid independent tests that has been demonstrated in the previous chapters is how the analyst determines the level of solution discrepancies. The reader should keep in mind that the best way for evaluating the accuracy of their particular problem is to compare the predictions with experimental measurements if available. Because the code verifications are performed for simpler cases where exact analytical solutions are known, the code accuracy may be limited to certain cases, etc. Therefore multiple methods for evaluating the accuracy are required for complex problems.

Moreover, some variables may convergence well, whereas other may be problematical or not convergence at all. Therefore while performing the grid convergence studies, critical variables should be examined carefully. Different discretisation schemes can be investigated as well.

## 9.5 Model Errors

These type of errors are obviously well understood by the reader. This may not only result due to geometrical data and input. Because the main issue here is the mathematical or empirical equations used to solve the problem. Therefore this type of problems will retain even after removing any kind of numerical errors. Hence, the analyst must understand the physical situation to be modelled and proceed with the model choice or development.

# References


[1] Inmon WH, Linstedt D. Front matter. Data Archit. a Prim. Data Sci. Elsevier; 2015. p. iii. Available from: https://doi.org/10.1016/B978-0-12-802044-9.00046-5.

[2] Meissner HC, Smith AL. The current status of chloramphenicol. Pediatrics 1979;64:348–56. Available from: https://doi.org/10.1016/0045-7949(79)90111-1.

[3] Attig N, Gibbon P, Lippert T. Trends in supercomputing: the European path to exascale. Comput Phys Commun 2011;182:2041–6. Available from: https://doi.org/10.1016/j.cpc.2010.11.011.

[4] http://www.top500.org; n.d.

[5] Taiji M, Narumi T, Ohno Y, Konagaya A. MDGRAPE-3: a petaflops special-purpose computer system for molecular dynamics simulations. Adv Parallel Comput 2004;13:669–76. Available from: https://doi.org/10.1016/S0927-5452(04)80083-2.

[6] Lim DJ, Anderson TR, Shott T. Technological forecasting of supercomputer development: The March to Exascale computing. Omega (United Kingdom) 2015;51:128–35. Available from: https://doi.org/10.1016/j.omega.2014.09.009.

[7] Fries TP. Higher-order conformal decomposition FEM (CDFEM). Comput Methods Appl Mech Eng 2017;328:75–98. Available from: https://doi.org/10.1016/j.cma.2017.08.046.

[8] Reuther A, Byun C, Arcand W, Bestor D, Bergeron B, Hubbell M, et al. Scalable system scheduling for HPC and big data. J Parallel Distrib Comput 2018;111:76–92. Available from: https://doi.org/10.1016/j.jpdc.2017.06.009.

[9] Cost efficient CFD simulations: proper selection of domain partitioning strategies. Comput Phys Commun 2017;219:121–34. Available from: https://doi.org/10.1016/J.CPC.2017.05.014.

[10] Paik SH, Moon JJ, Kim SJ, Lee M. Parallel performance of large scale impact simulations on Linux cluster super computer. Comput Struct 2006;84:732–41. Available from: https://doi.org/10.1016/j.compstruc.2005.11.013.

[11] Hua C, Cheng W-L. Energy saving evaluation of a novel energy system based on spray cooling for supercomputer center. Energy 2017. Available from: https://doi.org/10.1016/J.ENERGY.2017.09.089.

[12] Console P, Hairer E. Reducing round-off errors in symmetric multistep methods. J Comput Appl Math 2014;262:217–22. Available from: https://doi.org/10.1016/j.cam.2013.07.025.

[13] Vilmart G. Reducing round-off errors in rigid body dynamics. J Comput Phys 2008;227:7083–8. Available from: https://doi.org/10.1016/j.jcp.2008.04.013.

[14] Alvarez-Aramberri J, Pardo D, Paszynski M, Collier N, Dalcin L, Calo VM. On round-off error for adaptive finite element methods. Procedia Comput Sci 2012;9:1474–83. Available from: https://doi.org/10.1016/j.procs.2012.04.162.

# Index

*Note*: Page numbers followed by "*f*" and "*t*" refer to figures and tables, respectively.

N



O



P



R

Printed in the United States
By Bookmasters